THE MINKOWSKI MULTIDIMENSIONAL PROBLEM

SCRIPTA SERIES IN MATHEMATICS

Tikhonov and Arsenin · Solutions of Ill-Posed Problems, 1977

Rozanov · Innovation Processes, 1977

Pogorelov · The Minkowski Multidimensional Problem, 1978

Kolchin, Sevast'yanov, and Chist'yakov · Random Allocations, 1978

Boltyanskiy · Hilbert's Third Problem, 1978

THE MINKOWSKI MULTIDIMENSIONAL PROBLEM

ALEKSEY VASIL'YEVICH POGORELOV
USSR Academy of Sciences

Translated by Vladimir Oliker, University of Iowa and introduced by Louis Nirenberg, Courant Institute of Mathematical Sciences

1978
V. H. WINSTON & SONS
Washington, D.C.

A HALSTED PRESS BOOK

JOHN WILEY & SONS
New York Toronto London Sydney

V. H. Winston & Sons, a Division of Scripta Technica, Inc., Publishers
1511 K Street, N.W., Washington, D.C. 20005

Distributed solely by Halsted Press, a Division of John Wiley & Sons, Inc.

Library of Congress Cataloging in Publication Data

Pogorelov, Alekseĭ Vasil'evich.
The Minkowski multidimensional problem.

(Scripta series in mathematics)
Translation of Mnogomernaia͡ problema Minkovskogo.
Bibliography: p.
1. Convex surfaces. 2. Spaces, Generalized.
I. Title. II. Series.
QA643.P73813 516'.362 77–16680
ISBN 0-470-99358-8

Composition by **Isabelle Sneeringer**, Scripta Technica, Inc.

CONTENTS

INTRODUCTORY COMMENTARY

The Minkowski problem is that of finding a closed convex surface in R^3 whose Gaussian curvature is a given positive function K of the exterior unit normal; K necessarily satisfies the three conditions

$$\int \xi_i K^{-1}(\xi) d\omega(\xi) = 0 \quad i = 1, 2, 3$$

where integration is over the unit sphere.

In case K is a smooth function the problem has a smooth solution and a presentation of this result is contained in the, now classic, book [8] by the author. The Minkowski problem has a direct extension for closed convex hypersurfaces in R^n. However, after the result in R^3, it still took a number of years before the corresponding higher dimensional result was established–in a series of papers by A.V. Pogorelov (in the bibliography). This book presents a detailed treatment of the results described in those papers as well as generalizations of the Minkowski problem.

Analytically, the Minkowski problem involves solving a highly nonlinear partial differential equation of Monge-Ampère type. An example of such an equation, for a convex function $u(x^1, \ldots, x^n)$

defined in a domain in R^n is

$$\det\left(\frac{\partial^2 u}{\partial x^i \partial x^j}\right) = \phi(x) \tag{*}$$

where $\phi(x)$ is a given position function. To prove the existence of solutions of such equations, satisfying some boundary conditions, is no easy task. It involves finding *a priori* estimates for solutions and their derivatives up to third order. These estimates are all due to the author. In deriving bounds for the third derivatives he makes use of ideas of E. Calabi in [10]. This book also contains Pogorelov's proof of the conjecture (established previously for $n \leqslant 5$), that the only convex function satisfying (*) in all of R^n with $\phi = 1$ is a quadratic.

This book will introduce to the reader some beautiful geometric problems and a variety of interesting, deep, techniques for obtaining *a priori* estimates for elliptic equations–written by a master in the subject. It should serve as a basic reference for many years.

Attention of the reader should also be drawn to the recent papers by S. Y. Cheng and S. T. Yau on these topics:

On the regularity of the solution of the *n*-dimensional Minkowski problem, *Comm. Pure Appl. Math.*, (1976), **29**, pp. 495-516.

On the regularity of the Monge-Ampère equation

$$\det\left(\frac{\partial^2 u}{\partial x^i \partial x^j}\right) = F(x, u)$$

Comm. Pure Appl. Math., (1977), **30**, pp. 41-68.

Louis Nirenberg
Courant Institute of Mathematical Sciences

November 1977

PREFACE TO THE AMERICAN EDITION

Minkowski's problem is one whose solution has in the present century stimulated the development of the entire field of geometry, as well as of the theory of differential equations. Thus, a fundamental concept of the modern theory of differential equations, namely the concept of a generalized solution, first appeared in geometry in connection with Minkowski's own solution of his problem. Later, work on this problem has led to the development of general methods of proving the existence and uniqueness theorems in geometry and differential equation theory. Specifically, geometrical methods of analysis of elliptic equations of the Monge-Ampère type had been introduced in connection with the solution of the two-dimensional Minkowski problem.

The Minkowski problem and its generalizations have attracted much attention, and many papers of well-known American mathematicians, such as H. Lewy, J. Stoker, S. Chern, L. Nirenberg, and E. Calabi deal directly or indirectly with it. In particular, I should like to mention the papers by Nirenberg [6] and Calabi [10], which are cited extensively in the present book.

The translation of this book will enable English-speaking readers to become familiar more fully with the meaningful results, some of which had been previously discussed in the Doklady of the USSR Academy of Sciences.

A. Pogorelov
June 6, 1977

FOREWORD

An analytical approach to many problems in geometry leads to the study of partial differential equations. For example, the well-known classical Weyl and Minkowski problems in geometry "in the large" are related to the problem of solvability of the Monge-Ampère equation on a manifold homeomorphic to a sphere. During the past 25 years, significant achievements in the theory of Monge-Ampère elliptic differential equations have been made in conjunction with the two above-mentioned geometrical problems. In this way, mathematicians have obtained the general theorems on existence and uniqueness of solutions of Monge-Ampère equations. The geometric interpretation of an analytical problem suggested powerful geometric tools. One of these tools was the concept of a generalized solution. This concept was first applied to special types of equations of Weyl's and Minkowski's problems, but it was later used in the general case.

The equation

$$\det\left(\partial^2 z/\partial x_i \partial x_j\right) = \varphi(x_1, \ldots, x_n).$$

is a natural generalization of the simplest Monge-Ampère equation to the multidimensional case. A substantial part of this monograph is devoted to this equation. In particular, we define here a generalized solution, and we prove the solvability of the Dirichlet problem in any convex domain with continuous boundary conditions; the regularity of a strictly convex generalized solution when the right side (φ) of the equation is sufficiently regular is also proved. The study of all these questions is closely related to the multidimensional Minkowski problem, which, when analytically interpreted, leads to that equation. The regular solution of the multidimensional Minkowski problem is one of the principal results of the present work. This fact has determined its title.

INTRODUCTION

The present book is concerned with the well-known Minkowski problem as well as with a number of questions in geometry and differential equation theory that are related to this problem. The Minkowski problem is a problem of existence of a closed convex hypersurface whose Gaussian curvature is a given function with unit exterior normal. Minkowski [1] obtained a generalized solution of this problem. He proved that for a continuous and positive function $K(\xi)$ defined on a unit hypersphere Ω and satisfying the condition

$$\int_{\Omega} \xi \frac{d\omega}{K(\xi)} = 0,$$

there exists a closed convex hypersurface F whose Gaussian curvature at a point with exterior normal ξ is $K(\xi)$. This solution is called a generalized solution since it contains no information on the regularity of the hypersurface F even in the case in which $K(\xi)$ is an analytic function. Gaussian curvature is defined (following Gauss) as the limit of the ratio of the area of the spherical image to the area of a hypersurface and therefore, does not assume regularity of that hypersurface. If the given function $K(\xi)$ is a regular one, a natural question arises as to whether there exists a regular hypersurface with Gaussian curvature $K(\xi)$. This problem in the case of two-dimensional surfaces in three-dimensional space was studied by many well-known mathematicians: H. Lewy [4], C. Miranda [5], L. Nirenberg [6], and others.

In the series of papers [13]-[19], published by the author in *Doklady Akademii Nauk SSSR* [English translation *Soviet Math. Dokl.*, published by the American Mathematical Society], a solution of the general (multidimensional) Minkowski problem has been given. In these papers, some other questions in geometry and differential equations related to this problem were also considered. The present work contains a complete exposition of the results announced in the above-mentioned papers.

For the reader's convenience, we start with the basic definitions and results for convex bodies and hypersurfaces in E^n (Sec. 1). In the second paragraph, we reproduce the classic result of Minkowski. The third paragraph contains a regular solution of Minkowski's problem. It is proved there, that if $K(\xi)$ is positive and sufficiently regular, then the solution of Minkowski's problem is also regular. In particular, if $K(\xi)$ is an analytic function, then the solution is also analytic. The principal tool used in the proof is the *a priori* estimates of the principal radii of curvature of a hypersurface. In Sec. 4, estimates of this type have been obtained for hypersurfaces in the case in which an arbitrary function of the principal radii of curvature is given. The conditions imposed on this function are quite general. A natural generalization of Minkowski's problem is the question of existence of a closed convex hypersurface with preassigned curvature function of arbitrary order. This question is also studied in Sec. 4.

There is an important relationship between the solution of the Minkowski problem and the equation

$$\det(z_{ij}) = \varphi(x_1, \ldots, x_n).$$

The concept of a generalized solution of this equation is introduced and the Dirichlet problem is studied; it is also proved that, if the right side φ of the equation is sufficiently regular, then a strongly convex generalized solution of this equation is also regular.

The sixth paragraph is devoted to the improper affine hyperspheres. An improper affine hypersphere is a complete unbounded convex hypersurface $z = z(x_1, \ldots, x_n)$ satisfying the equation

$$\det (\zeta_{ij}) = \text{const} > 0.$$

It is proved that all improper affine hyperspheres are elliptic paraboloids. This result was obtained by K. Jörgens [9] for $n = 2$ and by E. Calabi [10] for $n = 3, 4, 5$. It is proved here for arbitrary n.

§1. CONVEX BODIES AND HYPERSURFACES IN E^n

In this paragraph, we shall recall the basic concepts and certain results for convex bodies and hypersurfaces in n-dimensional Euclidean space. A detailed exposition of these results with complete proofs can be found in the book by Bonnesen and Fenchel [2] and in Aleksandrov's papers [3].

1. Basic facts. A set M in a Euclidean space E^n is *convex* if it contains, with any two points x and y, the entire segment joining those two points. The intersection of two or more convex sets is convex. For any set H, the intersection of all convex sets containing H is called the *convex hull* of H. It coincides with the intersection of all the halfspaces containing H. When H is a convex set, it coincides with its convex hull.

A convex set V_a is called a *cone* with a vertex a if it consists of rays issuing from a point a. This point is called the *vertex of the cone.* Let T denote a convex set and let a denote a fixed point in a space. Then the rays issuing from the point a and intersecting the body T form a convex cone with the vertex a.

A convex set with interior points is called a *convex body*; a *convex hypersurface* is the boundary of a convex body. A convex hypersurface which bounds a bounded convex body is a manifold homeomorphic to a hypersphere. The homeomorphism can be established by projection from any interior point onto a hypersphere with that point as center. In the following, we shall usually consider bounded convex bodies and, correspondingly, closed hypersurfaces.

A convex polyhedron, which is the intersection of a finite number of halfspaces is a simple example of a nontrivial convex body. The boundary of a convex polyhedron consists of convex polyhedra of lower dimension and is also called a *polyhedron.* A convex polyhedron is the convex hull of a finite number of points which do not lie in a single hyperplane.

For convex bodies and the hypersurfaces bounding them, one introduces the concept of convergence. Specifically, a sequence of convex bodies T_n is said to converge to a body T if for every $\varepsilon > 0$ there exists a sufficiently large N such that the body T_n is contained in the ε-neighborhood of the body T and the body T is contained in the ε-neighborhood of the body T_n. Other equivalent definitions of convergence for convex bodies can be given. In particular, let O denote an interior point of a body T and let τ be an arbitrary ray issuing from O. Let x and x_n denote the points of intersection of τ and the boundaries of T and T_n, respectively. Then the sequence of bodies T_n converges to the body T if, for every τ, $x_n \to x$ as $n \to \infty$. If a bounded sequence of convex bodies contains a ball, then this sequence contains a subsequence which converges to a convex body T.

In the theory of convex bodies and hypersurfaces, the concept of a support hyperplane is of great importance. Let x denote an arbitrary point of the convex hypersurface F which bounds a body T. A hyperplane α passing through the point x is a *support hyperplane* of the body T (hypersurface F) at the point x if all points of the body lie on one side of the hyperplane α. There is at least one support hyperplane passing through each point of a convex hypersurface. If there is only one support hyperplane at the point x, then x is called a *smooth point.* A hypersurface is *smooth* when all its points are smooth. A hypersurface is called *strongly convex* if each of its support hyperplanes has only one common point with this hypersurface.

Let us indicate some properties of support hyperplanes. Let F be a convex hypersurface, x_n a sequence of points on F converging to a point x_0, and α_n a support hyperplane at x_n. Then if the sequence of hyperplanes α_n converges to a hyperplane α_0, the latter is a support hyperplane of F at x_0. Let F_n denote a sequence of convex hypersurfaces converging to a hypersurface F, let x_n denote a point on F_n, and let α_n denote a support hyperplane to F at the point x_n. Then if the point sequence x_n converges to a point x_0 on F and the sequence of hyperplanes α_n converges to a hyperplane α_0, the latter is a support hyperplane of F at the point x_0.

Another important concept for convex bodies and convex hypersurfaces is the concept of a support function. Let F denote a hypersurface bounding a convex body T in E^n. We shall introduce Cartesian coordinates $x_1, \ldots, x_n$ with the origin inside T. A *support function* of the body T (hypersurface F) is a function $H(x)$ defined on E^n by

$$H(x) = \sup_{y \subset T} (xy),$$

The function $H(x)$ is positively homogeneous of first degree, and convex. This means that for $\lambda \geqslant 0$

$$H(\lambda\, x) = \lambda\, H(x),$$

and for $\lambda \geqslant 0$, $\mu \geqslant 0$

$$H(\lambda x + \mu y) \leqslant \lambda H(x) + \mu H(y).$$

Conversely, a convex function $H(x)$ that is positive homogeneous of first degree is a support function of a certain convex body (possibly a degenerate one). This body is the intersection of the half-spaces E_y defined by the inequalities

$$(xy) \leqslant H\,(y).$$

If a convex body is translated along a vector a, the support function is modified by addition of a linear term ax. A support function has a simple meaning: for a unit vector x, a support function is the distance from the coordinate origin to the support hyperplane with exterior normal x; in the general case, this distance is multiplied by $|x|$.

2. Regular convex hypersurfaces. A convex hypersurface in E^n is called *regular* if it admits a regular parametrization in a neighborhood of an arbitrary point; i.e., the hypersurface can be defined by a regular* vector function

$$x = x(u_1, \ldots, u_{n-1}),$$

such that the quadratic differential form $(dx)^2$ is positive-definite. We shall say that a hypersurface F is of class $C^{k,\alpha}$, where $k \geqslant 1$ and $0 < \alpha < 1$, if there exists a regular local parametrization of class $C^{k,\alpha}$, i.e., if the vector function $x(u)$ defining the hypersurface is k times continuously differentiable and its partial derivatives satisfy the Hölder condition with exponent α. If a hypersurface admits a regular parametrization of class $C^{k,\alpha}$, where $k \geqslant 1$, then in a neighborhood of an arbitrary point in suitably chosen Cartesian coordinates it admits an explicit representation by the equation

$$x_n = \varphi(x_1, \ldots, x_{n-1}),$$

where φ is a function of class $C^{k,\alpha}$. In particular, to define in such a way a hypersurface in the neighborhood of an arbitrary point A, it suffices to take the support hyperplane at the point A as coordinate hyperplane $x_1, x_2, \ldots, x_{n-1}$. A regular hypersurface ($k \geqslant 1$) is smooth; its support hyperplane at an arbitrary point $x(u)$ is parallel to the vectors $\partial x/\partial u_1, \partial x/\partial u_2, \ldots, \partial x/\partial u_{n-1}$. In the case of explicit representation of the hypersurface by an equation $x_n = \varphi(x_1, \ldots, x_{n-1})$, the support (tangent) hyperplane has slopes $\partial\varphi/\partial x_1, \partial\varphi/\partial x_2, \ldots, \partial\varphi/\partial x_{n-1}$.

Consider an arbitrary point O on a regular convex hypersurface F in E^{n+1} and a support (tangent) hyperplane α passing through O. Let d denote an arbitrary direction from the point O in the hyperplane α. The two-dimensional plane determined by the normal of the hypersurface at the point O and the direction d cuts out of the hypersurface a curve γ_d, called a *normal section.* The

*Editor's note. Of class C^k, where $k \geqslant 1$.

curvature of γ_d at the point O is called the *normal curvature* of the hypersurface in the direction d. Let us introduce in E^{n+1} Cartesian coordinates $x_0, x_1, x_2, \ldots, x_n$, assuming that the coordinate hyperplane $x_1, \ldots, x_n$ is the hyperplane α. In the neighborhood of the point O, the hypersurface F is defined by the equation

$$x_0 = \varphi(x_1, x_2, \ldots, x_n),$$

and its normal curvature at the point O in the direction d is

$$k_d = d^2\varphi/(dx)^2.$$

For a convex hypersurface, the normal curvature in any direction d is nonnegative: $k_d \geqslant 0$. A direction d from the point O on the hypersurface is called the *principal* direction if the normal curvature k_d is stationary in that direction. For the direction of the x_1 axis at the point O to be the principal direction it is necessary and sufficient that $\partial^2\varphi/\partial x_k \partial x_1 = 0$ for $k > 1$, and $\partial^2\varphi/\partial x_1^2 = k_1$.

Let r denote the position vector of a point on a hypersurface and let ν denote the unit vector of the exterior normal at this point. If the direction d is the principal direction, then we have in this direction

$$k_d dr = d\nu \quad \text{(Rodrigues' formula).}$$

If a convex hypersurface is regular and at least twice differentiable, and if at any point and in any direction its normal curvature is strictly positive, then the support function is of the same regularity class.* In this case, the first derivatives of the support function $\partial H/\partial x_i$ have a simple meaning: they are the coordinates of the point at which the support hyperplane with exterior normal ν is tangent to the hyperstructure. We shall use this property to define the principal radii of curvature of the hypersurface in terms of the support function. Let x denote the position vector of an arbitrary point of the hypersurface and let ξ

Editor's note. See editor's note on page 6.

denote the unit exterior normal at this point. For the principal direction according to the formula of Rodrigues,

$$dx - R d\xi = 0.$$

Noting that $x = \nabla H(\xi)$, we obtain

$$\sum_j \left(\frac{\partial^2 H}{\partial \xi_i \, \partial \xi_j} d\xi_j - R \, d\xi_i \right) = 0.$$

Hence, the principal radii of curvature satisfy the equation

$$\det \left(\frac{\partial^2 H}{\partial \xi_i \, \partial \xi_j} - R\delta_{ij} \right) = 0.$$

Since H is homogeneous and $\det(\partial^2 H/\partial \xi_i \partial \xi_j) = 0$, one root of this equation is $R = 0$. Disregarding this unessential value of R, we obtain for other values of R an equation of the form

$$R^n - A_1 R^{n-1} + \ldots + A_n = 0,$$

where the A_i are the sums of the principal ith order minors of the matrix $(\partial^2 H / \partial \xi_i \partial \xi_j)$ on the unit sphere $|\xi| = 1$.

For a local study of a hypersurface in a neighborhood of a given point A, it is convenient to choose Cartesian coordinates $x_0, x_1, \ldots, x_n$ in the following way: We place the origin inside the body bounded by the hypersurface and on the interior normal at the point A; the coordinate hyperplane $x_1, x_2, \ldots, x_n$ is taken parallel to the tangent hyperplane at the point A, and the directions of the coordinate axes are such that $x_1, x_2, \ldots, x_n$ are parallel to the principal directions of the hypersurface at the point A. Instead of the support function H, we introduce the function

$$h(x_1, \ldots, x_n) = H(1, x_1, \ldots, x_n).$$

In this coordinate system, we have at the point A: $\partial h/\partial x_i = 0$, $\partial^2 h/\partial x_i \partial x_j = 0$ for $i \neq j$, $\partial^2 h/\partial x_i^2 = R_i$, where R_i is the principal radius of curvature of the hypersurface in the direction of the x_i axis.

Let us find the equation for the principal radii of curvature of the hypersurface at a point with exterior normal of direction $(1, x_1, x_2, \ldots, x_n)$. We have

$$H(x_0, x_1, \ldots, x_n) = x_0 h(x_1/x_0, \ldots, x_n/x_0).$$

Differentiating this equation, we obtain successively

$$\frac{\partial H}{\partial x_k} = h_k, \quad k > 0, \qquad \frac{\partial H}{\partial x_0} = h - \sum_{\alpha>0} \frac{x_\alpha}{x_0} h_\alpha,$$

$$\frac{\partial^2 H}{\partial x_i\, \partial x_j} = \frac{1}{x_0} h_{ij} \qquad \text{for} \quad i, j > 0,$$

$$\frac{\partial^2 H}{\partial x_0\, \partial x_k} = -\sum_{\alpha>0} \frac{x_\alpha}{x_0^2} h_{\alpha k} \qquad \text{for} \quad k > 0.$$

$$\frac{\partial^2 H}{\partial x_0^2} = \sum_{\alpha, \beta>0} \frac{x_\alpha x_\beta}{x_0^3} h_{\alpha\beta}.$$

If we substitute these values of the derivatives into the equation $\det(H_{ij} - R\delta_{ij}) = 0$ and simplify the first row and the first column of the determinant by means of the other rows and columns, the equation takes the form

$$\begin{Vmatrix} -\frac{R}{x_0^2}\left(x_0^2 + \sum_{\alpha>0} x_\alpha^2\right) & R\frac{x_1}{x_0} & \ldots & R\frac{x_n}{x_0} \\ R\frac{x_1}{x_0} & \frac{h_{11}}{x_0} - R & \ldots & \frac{h_{1n}}{x_0} \\ \ldots & \ldots & \ldots & \ldots \\ R\frac{x_n}{x_0} & \frac{h_{n1}}{x_0} & \ldots & \frac{h_{nn}}{x_0} - R \end{Vmatrix} = 0.$$

In this equation, we substitute the coordinates of the exterior normal vector of direction $1, x_1, \ldots, x_n$, i.e., the values

$$\frac{1}{\left(1+\sum\limits_{\alpha>0} x_\alpha^2\right)^{1/2}}, \frac{x_1}{\left(1+\sum\limits_{\alpha>0} x_\alpha^2\right)^{1/2}}, \ldots, \frac{x_n}{\left(1+\sum\limits_{\alpha>0} x_\alpha^2\right)^{1/2}}.$$

As a result, the equation for the principal radii of curvature at the point with this normal will be

$$\left\| \begin{matrix} -R\lambda^2 & Rx_1 & Rx_2 & \ldots & Rx_n \\ Rx_1 & \lambda h_{11}-R & \lambda h_{12} & \ldots & \lambda h_{1n} \\ \cdot & \cdot & \cdot & \cdot & \cdot \\ Rx_n & \lambda h_{n1} & \lambda h_{n2} & \ldots & \lambda h_{nn}-R \end{matrix} \right\| = 0,$$

where $\lambda = (1 + x_1^2 + \ldots + x_n^2)^{1/2}$.

3. Area and curvature of a convex hypersurface. Let F denote a convex hypersurface. We mark a point O inside the body bounded by the hypersurface. Let us consider a sequence of polyhedra P_n converging to the hypersurface F. When n is sufficiently large, the point O is interior to the polyhedra P_n. Let M denote a Borel set on the hypersurface F. Project the set M onto the polyhedron P_n from the center O. We denote by $S_{n\alpha}$ the area (*Lebesgue measure*) of the projection of the set M onto the face α of the polyhedron P_n. Let us define

$$S_n = \sum_\alpha S_{n\alpha},$$

where the summation is over all faces α of the polyhedron P_n. It turns out that, as $n \to \infty$, the sequence S_n converges to a limit $S(M)$, which depends neither on the choice of the point O nor on the sequence of polyhedra P_n. This limit $S(M)$ is called the *area of the hypersurface on the set M.*

For a regular hypersurface defined on the set M by a vector-function $r(u_1, \ldots, u_n)$, this area can be computed by the formula

$$S(M) = \int_M \sqrt{g}\, du_1 \ldots du_n,$$

where g is the discriminant of the quadratic form $(dr)^2$. In case of an explicit representation of the hypersurface by the equation $z = z(x_1, \ldots, x_n)$, we have for $S(M)$:

$$S(M) = \int_M \sqrt{1 + (\nabla z)^2}\, dx_1 \ldots dx_n.$$

Let M denote an arbitrary set on the hypersurface F. We shall define a set M^* on a unit hypersphere Ω in the following way. A point x on Ω belongs to the set M^* if there exists a point x' in M and a support hyperplane α' at x' such that the exterior normal to α' translated to the center of Ω has x as its endpoint. This set M^* is called the *spherical image* of the set M. If the set M is a Borel set, its spherical image is also a Borel set and, therefore, is measurable. In this case, the area of the set M^* on the hypersphere Ω is called the *curvature of the set M on the hypersurface F.* The curvature, being a set function, is a nonnegative, completely additive function on the set of Borel sets. It possesses an important convergence property. Specifically, when a sequence of convex hypersurfaces converges to the hypersurface F, their curvatures converge weakly to the curvature of F. In particular, if one projects the hypersurfaces F_n on the hypersurface F from an interior point, then for any continuous function f defined on F,

$$\lim \int_F f\, d\omega_n = \int_F f\, d\omega.$$

Projection here reduces to transference of the function f, defined on F, onto the hypersurface F_n.

Let x denote an arbitrary point and G an arbitrary domain on a hypersurface F. Denote the area of the domain G by $S(G)$ and its curvature by $\omega(G)$. The limit, if it exists, of the ratio $\omega(G)/S(G)$, as the domain G shrinks to the point x is called the *Gaussian curvature of the hypersurface at the point x*. For a regular twice differentiable convex hypersurface, the Gaussian curvature is defined at each point and is equal to the product of the principal curvatures. Indeed, since it is always possible to parametrize the hypersurface so that the directions of the coordinate curves (u) at the point x are the principal directions, we have by Rodrigues' formula $k_\alpha r_\alpha = \nu_\alpha$. The area of the domain G is

$$S(G) = \int_G \sqrt{g_r}\, du_1 \ldots du_n.$$

The area of the spherical image of the domain is

$$\omega(G) = \int_G \sqrt{g_\nu}\, du_1 \ldots du_n.$$

As the domain G shrinks to the point x, the ratio $\sqrt{g_\nu}/\sqrt{g_r}$ converges to the limit $k_1 k_2 \ldots k_n$. Hence $\omega(G)/S(G)$ converges to the same limit, which is the product of the principal curvatures. It is worth noting that the existence or even the continuity of the Gaussian curvature in terms of the above-given general definition does not imply that the hypersurface is of class C^2. (A. D. Aleksandrov).

As was shown in subsection 2, the principal radii of curvature of a convex hypersurface are the roots of the equation

$$\frac{1}{R} \det (H_{\alpha\beta} - R\delta_{\alpha\beta}) = 0,$$

where the derivatives $H_{\alpha\beta}$ are taken on the unit hypersphere. The free term in this equation is the product of the principal radii of curvature and it is the sum of the principal minors of $\det (H_{\alpha\beta})$. Its reciprocal is the Gaussian curvature, which is defined in terms of $h(x_1, \ldots, x_n) = H(1, x_1, \ldots, x_n)$ by

$$K = \frac{1}{(h_{\alpha\beta})(1 + x_1^2 + \ldots + x_n^2)^{\frac{n}{2}+1}}.$$

Let M denote an arbitrary point set on a unit hypersphere. A set M' on a convex hypersurface is defined as a set such that, for each point $x \in M'$, the spherical image of x belongs to M. It turns out that, if M is a Borel set, M' is also a Borel set. In this case, its area is defined and we denote it by $\sigma(M)$. The quantity σ is a set function, known as the *surface area function.* The surface area function, defined by means of a convex hypersurface, proves to be a nonnegative, completely additive function on the ring of Borel sets of the unit hypersphere. Like curvature, the surface area functions of convex hypersurfaces F_n converging to a hypersurface F converge weakly to the surface area function of the hypersurface F. It means that, for each continuous function f defined on a unit hypersphere Ω,

$$\lim \int_{\Omega} f\, d\sigma_n = \int_{\Omega} f\, d\sigma,$$

where σ_n and σ denote the surface area functions of F_n and F.

For a regular twice differentiable convex hypersurface F, the surface area function admits a simple analytic representation:

$$\sigma(M) = \int_{M} \frac{d\omega_\xi}{K(\xi)},$$

where $K(\xi)$ is the Gaussian curvature of the hypersurface at the point with exterior normal ξ, $d\omega_\xi$ is the area element, and the domain of integration is the set M.

The volume of the body bounded by a convex hypersurface can easily be defined by means of the surface area function, specifically,

$$V(T) = \frac{1}{n} \int_{F} H\, d\sigma,$$

where H is the support function of the body T and σ is the surface area function. This formula is obvious when the body is a polyhedron. In the general case, it can be obtained by approximating the body T with a convex polyhedron P and taking the limit as $P \to T$. For bodies with regular boundaries, it was shown above that $d\sigma = R_1 \ldots R_n d\omega$. The product of the principal radii of curvature $R_1 R_2 \ldots R_n = \Sigma \, |H_{\alpha\beta}|'$, where the $|H_{\alpha\beta}|'$ are the principal minors of det $(H_{\alpha\beta})$ on a unit sphere. We thus obtain the important formula

$$V = \frac{1}{n} \int_{\Omega} H \sum |H_{\alpha\beta}|' \, d\omega.$$

4. Mixture of convex bodies. For convex bodies $T_1, T_2, \ldots, T_m$ and nonnegative real numbers $\lambda_1, \lambda_2, \ldots, \lambda_m$, we consider a body T consisting of the points

$$x = \lambda_1 x_1 + \lambda_2 x_2 + \ldots + \lambda_m x_m ,$$

where $x_k \in T_k$, $k = 1, 2, \ldots, m$. The convexity of the bodies T_k implies that T is also convex. The body T is said to be a *mixture* of the bodies T_k. The mixture of convex bodies was first considered by Brunn and Minkowski [1]. It follows from the definition of the support function of a convex body that the support function of the body T is

$$H = \sum_k \lambda_k H_k,$$

where the H_k are the support functions of the bodies T_k.

In terms of support and surface area functions, the volume of the body T can be expressed as

$$V = \frac{1}{n} \int H \sum |H_{\alpha\beta}|' \, d\omega$$

(see subsection 3)

Now, if T is the mixture of the convex bodies T_k, $k = 1, \ldots, m$, with regular boundaries, then one easily shows that the volume of T is a homogeneous polynomial in the variables λ_k, where the coefficients depend only on the bodies T_k, and

$$V(T) = \sum \lambda_{k_1} \lambda_{k_2} \ldots \lambda_{k_m} V(T_{k_1}, T_{k_2}, \ldots, T_{k_m}).$$

The coefficients $V(T_{k_1}, \ldots, T_{k_m})$ are called the *mixed volumes.* This result is extended to arbitrary convex bodies by approximation with bodies with regular boundaries and a limit process.

From the expression for the volume of the body T in terms of the surface area function, one obtains the following formula for the mixed volume:

$$V(T_1, T_2, \ldots, T_2) = \frac{1}{n} \int_{\Omega} H_2 \, d\sigma_1,$$

where H_2 is the support function of the body T_2 and σ_1 is the surface area function of the body T_1.

Of special interest is the mixture of two bodies T_0 and T_1:

$$T_\vartheta = (1 - \vartheta) T_0 + \vartheta T_1 \qquad (0 \leqslant \vartheta \leqslant 1).$$

The function $\sqrt[n]{V(T_\vartheta)}$ is concave with respect to ϑ:

$$\sqrt[n]{V(T_\vartheta)} \geqslant (1 - \vartheta) \sqrt[n]{V(T_0)} + \vartheta \sqrt[n]{V(T_1)},$$

with equality holding only when T_0 and T_1 are homothetic (*The Brunn-Minkowski theorem*). The fact that the function $\sqrt[n]{V(T_\vartheta)}$ is concave with respect to ϑ leads to the *Minkowski inequality* for mixed volumes:

$$V^n(T_0, T_1, \ldots, T_1) \geqslant V^{n-1}(T_1) V(T_0),$$

with equality holding only when T_0 and T_1 are homothetic.

For a given hypersurface, we denote by S_m the normalized elementary symmetric function of the principal radii of curvature:

$$S_m = \frac{1}{\binom{n}{m}} \sum_{\alpha_i \neq \alpha_j} R_{\alpha_1} R_{\alpha_2} \ldots R_{\alpha_m}$$

We shall call S_m the *curvature function of order m*. It is known that it can be expressed in terms of the sum of the principal minors of order m of the matrix $(H_{\alpha\beta})$. Hence, for $T = \lambda_1 T_1 + \lambda_2 T_2$ the curvature function S_m is a homogeneous polynomial of degree m in λ_1 and λ_2. Its coefficients are called the *mixed curvature functions.*

For mixed curvature functions, A. D. Aleksandrov [3] proved that

$$[S_m (T_1, T_2, \ldots, T_2)]^m \geqslant S_m (T_1, \ldots, T_1) [S_m (T_2, \ldots, T_2)]^{m-1}.$$

with equality holding here if and only if the quadratic forms $d^2 H_1$ and $d^2 H_2$ are proportional.

Finally, let us note the following relationship between mixed curvature functions and mixed volumes [3]:

$$\int_\Omega S_m (T_1, \underbrace{T_2, \ldots, T_2}_{m-1}) H_2 \, d\omega = V (T_1, \underbrace{T_2, \ldots, T_2}_{m}, E, \ldots, E),$$

$$\int_\Omega S_m (T_1, T_1, \ldots, T_1) H_2 \, d\omega = V (T_2, T_1, \ldots, T_1, E, \ldots, E).$$

In these formulas, E denotes a unit ball.

§2. GENERALIZED SOLUTION OF THE MINKOWSKI PROBLEM

In this section, we present the classical result of Minkowski on the problem of existence and uniqueness of a convex hypersurface with preassigned Gaussian curvature $K(\nu)$. The solution of this problem posed by Minkowski is called a generalized solution since the result does not say anything about regularity of the solution, even in the case in which the given function $K(\nu)$ is analytic. The question as to when the solution of Minkowski's problem is a regular one, provided $K(\nu)$ is sufficiently regular, will be answered in the next section.

1. Statement of the problem. Uniqueness of the solution. Let $K(\nu)$ denote a positive continuous function defined on the hypersphere Ω. Minkowski's problem is the problem of existence of a convex hypersurface F with Gaussian curvature $K(\nu)$ at a point with exterior normal ν. The Gaussian curvature is understood in the sense of the definition given in Sec. 1, subsection 2. Specifically, for a given hypersurface F and a point $x \in F$, let us consider the ratio $\omega(G)/S(G)$, where $S(G)$ is the area of a domain G containing x and $\omega(G)$ is the spherical image of G. The *Gaussian curvature of F at x* is the limit of this ratio as G shrinks to the point x. Such a definition does not assume regularity of the hypersurface F.

If the curvature of a hypersurface is given as a function of the exterior normal, then the surface area function is defined and

$$\sigma = \int \frac{d\omega}{K(\nu)} \cdot$$

Conversely, defining the surface area function as a set function on a unit hypersphere is the same as defining the Gaussian curvature as a function of the exterior normal. Hence, the problem can naturally be generalized to the problem of finding conditions for a nonnegative, completely additive set function σ defined on a unit hypersphere to be a surface area function of some convex hypersurface. In particular, when the function σ is equal to zero everywhere except at a finite number of points $\nu_1, \ldots, \nu_m$, at which it assumes positive values $\sigma_1, \ldots, \sigma_m$, we have the problem of existence of a convex polyhedron with faces of areas σ_k and with exterior normals at ν_k.

Minkowski solved the problem first for convex polyhedron. Then, for a given function $K(\nu)$, he constructs a set function σ on the hypersphere. This function σ is approximated by a function σ' that is zero everywhere except at a finite number of points ν_k, at which it is equal to σ_k. To find the function σ', one partitions the hypersphere into small domains g_k. Then the vectors ν_k and the numbers σ_k are determined by the condition

$$\sigma_k \nu_k = \int_{g_k} \nu \, d\sigma.$$

The next step is to prove the existence of a polyhedron P_g with faces of area σ_k and normals ν_k. Finally, letting the diameters of g_k tend to zero, we obtain a sequence of polyhedra that converges to a convex hypersurface. This hypersurface is the solution of the problem; i.e., it has the preassigned curvature $K(\nu)$ at each point with exterior normal ν.

When a hypersurface F is a solution of the problem, the hypersurface F', which is obtained by translation of F, is also a solution. Thus, a certain necessary integrability condition should be imposed on the surface area function. At first, we consider the case of a polyhedron. Suppose that a polyhedron P' is obtained from P by translation by a small distance. One may think of this translation as a deformation of P. Let ν_k denote the exterior normals of P and S_k the areas of the faces. Suppose that P is translated a distance ε in the direction of e. Displacement of a face with normal ν_k causes a change in the volume $\simeq S_k(\nu_k e)\,\varepsilon$. Since P is equal to P', the total change in the volume is zero. Therefore,

$$\sum_k (\nu_k e)\, S_k = 0.$$

The vector e was chosen arbitrarily. Hence,

$$\sum_k \nu_k S_k = 0.$$

Let us now consider the general case. For a given convex hypersurface F, we construct a sequence of convex polyhedra P_n converging to F. The surface area functions of P_n converge weakly to the surface area function of F. For the polyhedra,

$$\int \nu \, d\sigma_n = 0.$$

For this reason and because of the weak convergence of the surface area function, it follows for F that

$$\int_{\Omega} \nu \, d\sigma = 0.$$

Thus, this condition is a necessary one for existence of a convex hypersurface with a given surface area function σ. If the hypersurface has a Gaussian curvature, then $d\sigma = d\omega / K(\nu)$ and the necessary condition can be written

$$\int_{\Omega} \frac{\nu \, d\omega}{K(\nu)} = 0.$$

Let us now consider the uniqueness problem. We shall show that, if two convex hypersurfaces F_1 and F_2 have the same surface area function, then they are equal up to translation. Let T_1 and T_2 denote two convex bodies bounded by F_1 and F_2, let H_1 and H_2 denote their support functions, and let σ_1 and σ_2 denote the surface area functions of F_1 and F_2. We have (see subsection 4, Sec. 1)

$$V(T_1, T_2, \ldots, T_2) = \frac{1}{n} \int_{\Omega} H_2 \, d\sigma_1.$$

Since $\sigma_1 \equiv \sigma_2$, the right-hand side of this equation is the volume of T_2. Thus,

$$V(T_1, T_2, \ldots, T_2) = V(T_2).$$

By Minkowski's inequality for mixed volumes,

$$V^n(T_1, T_2, \ldots, T_2) \geqslant V^{n-1}(T_2) V(T_1).$$

This inequality and the previous equality now yield the result that $V(T_2) \geqslant V(T_1)$. Interchanging T_1 and T_2, we obtain $V(T_1) \geqslant V(T_2)$. Hence, $V(T_1) = V(T_2)$. This fact implies that equality holds in Minkowski's inequality. This is possible if and only if F_1 and F_2 differ by translation. The uniqueness is proved. Thus, a convex hypersurface is uniquely determined, up to translation, by its surface area function. In particular, a convex polyhedron is uniquely determined by the areas of its faces and their exterior normals.

2. Existence of a convex polyhedron with faces of preassigned area and given exterior normals. Let $\xi_1, \xi_2, \ldots, \xi_m$ denote a set of noncoplanar unit vectors and let $\sigma_1, \sigma_2, \ldots, \sigma_m$ denote positive numbers such that

$$\sum_k \sigma_k \xi_k = 0. \tag{1}$$

Then there exists a convex polyhedron with faces of areas σ_k and exterior normals ξ_k. This polyhedron is unique up to translation (*Minkowski's theorem*).

The proof of this theorem will be based on A. D. Aleksandrov's "mapping lemma." Let A and B denote two manifolds of the same dimension. Let φ be a mapping of A into B with the following properties:

1. Each component of B contains images of points of A.
2. The mapping φ is injective and continuous.
3. If a sequence of points b_k of B, which are images of points a_k of A converges to a point b, then A contains a point a whose image is b and which is a limit point of the points a_k.*

To make use of this lemma in the proof of Minkowski's theorem, let us note some properties of polyhedra satisfying condition (1). First note that exterior normals ξ_k cannot point into the same halfspace. Suppose that this is not true. Then, since the vectors ξ_k

*Editor's note. The mapping lemma reads: When conditions 1, 2, and 3 are satisfied, then $\varphi(A) = B$, i.e., φ maps A onto B. (see page 23) are noncoplanar and the numbers σ_k are positive, the vector $\Sigma\sigma_k\xi_k$ is a nonzero vector and this contradicts condition (1).

For any unit vectors ξ_k, for $k = 1, \ldots, m$, satisfying condition (1), there exists a convex polyhedron with faces whose exterior normals are ξ_k. To prove this, take a unit hypersphere Ω and consider the tangent hyperplanes to Ω at the endpoints of the vectors ξ_k issuing from the center of Ω. For each hypersurface, mark the half-space containing the hypersphere Ω. The intersection of the half-spaces is a convex polyhedron P. The polyhedron P must be bounded. For if P were unbounded, it would contain a ray l issuing from the center of Ω and hence the vectors ξ_k would all point into the halfspace defined by the hyperplane passing through the center of Ω and perpendicular to l.

We assert that the set of all convex polyhedra with exterior normals ξ_k such that surface areas of their faces σ_k satisfy the condition $0 < a \leqslant \sigma_k \leqslant b < \infty$ is bounded, i.e., that they are all contained with a ball of sufficiently large radius. Suppose the assertion is not true. Then among the polyhedra in question there exist polyhedra of arbitrary large diameters. Let P denote such a polyhedron with diameter D and let A and B denote points such that $AB = D$. Let us show that the projection of this polyhedron onto the hyperplane perpendicular to the line AB has an arbitrarily small area when D is sufficiently large. Let us perform Steiner's symmetrization of the polyhedron P with respect to a hyperplane α through the midpoint of AB and perpendicular to AB. We obtain a new polyhedron P', which is symmetric to P with respect to the hyperplane α and has the same volume and the same area of the projection on α. The projection of P' onto the hyperplane α is a section of the polyhedron P' by α. Denote this section by B'. The two cones with vertices A and B and common base B' are contained in P'; hence the sum of their volumes is no larger than the volume V of the polyhedron P'. Thus, the following inequality is obtained

$$\frac{1}{n} DS \leqslant V,$$

where S is the area of B'. Since the polyhedra in question have a uniformly bounded surface area no greater then mb), they have, by the isoperimetric inequality, a uniformly bounded volume. Hence, the projection of a polyhedron onto a hyperplane perpendicular to the line AB has an arbitrarily small area S if the diameter of the polyhedron is sufficiently great. But since the vectors ξ_k are noncoplanar and the surface areas of the faces are bounded from below ($\sigma_k \geqslant a > 0$), the area of the projection of the polyhedron P onto any hyperplane is bounded from below by a positive number. This contradiction proves our assertion.

Let us now prove the theorem. We first define the manifolds A and B from the mapping lemma. The points of A are classes of translation-equivalent polyhedra. Since each polyhedron is determined by its support numbers (the distances of face hyperplanes from a certain fixed interior point) and there are m faces, the manifold A is of dimension $m - n$. The points of B are the points of m-dimensional Euclidean space with coordinates σ_k such that $\sigma_k > 0$ and $\Sigma\sigma_k\xi_k = 0$. The dimension of B is also $m - n$. The mapping φ in the lemma assigns to each class of equivalent polyhedra a point of B with coordinates $\sigma_1, \ldots, \sigma_m$ equal to the appropriate areas of the faces of a polyhedron of this class. Let us show that the assumptions of the mapping lemma hold for the manifolds A and B and the mapping φ.

The first condition of the mapping lemma holds since the manifold B (being the intersection of convex sets) is convex and hence, connected. As was shown above, there exists a polyhedron with faces whose exterior normals are ξ_k. Hence, the pre-image of B is a nonempty subset of A. The second condition requires that the mapping φ be one-to-one. That this condition holds follows from the uniqueness result proved in subsection 1 of this section. Obviously, φ is continuous.

Finally, let us verify the third condition of the lemma. In the present case, we consider a sequence of polyhedra P^k such that the areas σ_α^k of their faces converge to numbers $\sigma_\alpha > 0$. We have to show that this sequence of polyhedra, after being suitably translated, is convergent and that the areas of the limit polyhedron are σ_α. Let the origin O of the coordinate system be the center of gravity of the polyhedra P^k, which can always be attained by means of a translation. By Blaschke's theorem, the sequence of the polyhedra P^k contains a subsequence converging to a convex body which might, at first, be degenerate. But degeneracy is impossible since it means that, when k is sufficiently large, the area of the projection of P^k onto a certain hyperplane is arbitrarily small. Hence, the limit body P is not degenerate. It is obvious that P is a polyhedron with faces of area σ_α and exterior normal ξ_α; the center of gravity is at the origin O of the coordinate system. Any other convergent subsequence P^k has its limit polyhedron with the same exterior normals, areas and center of gravity and therefore, coincides with P. Therefore, the third condition of the lemma also holds.

By the mapping lemma, φ is an onto mapping, i.e., for any system of positive numbers σ_k satisfying the condition $\Sigma\sigma_k\xi_k = 0$, there exists a polyhedron with faces of area σ_k and exterior normals ξ_k. The theorem is proved.

3. Existence of a convex hypersurface with given Gaussian curvature. Let Ω denote a unit hypersphere and let $K(\xi)$ denote a given continuous positive function satisfying the condition

$$\int_\Omega \frac{\xi d\omega}{K(\xi)} = 0.$$

Then there exists a convex hypersurface F, unique up to translation, for which $K(\xi)$ is the Gaussian curvature at the point with exterior normal ξ (*Minkowski's theorem*).

PROOF. We decompose the hypersphere Ω into small domains g_k. Define the numbers σ_k and unit vectors ξ_k by

$$\sigma_k\xi_k = \int_{g_k} \frac{\xi d\omega}{K(\xi)}.$$

Obviously

$$\sum \sigma_k\xi_k = 0.$$

Hence, there exists a convex polyhedron P with faces of area σ_k and exterior normals ξ_k.

Construct a sequence of decompositions of the hypersphere such that the diameters of the domains g_k in the mth decomposition approach zero as $m \to \infty$. Let P^m denote a convex polyhedron constructed for the mth decomposition of the hypersphere into domains $\overset{m}{g_k}$. We shall show that the polyhedra P^m are uniformly bounded. Let d_m denote the diameter of the polyhedron P^m and let A and B denote two points the distance between which is d_m. We denote by α the hyperplane through the midpoint of AB perpendicular to AB. Let Q^m denote the projection of the polyhedron P^m onto the hyperplane α and let S^m denote the area of Q^m.

Let us show that S^m is bounded from below by a certain number S^0. We have

$$S^m = \frac{1}{2} \sum_k \sigma_k \, |\xi_k e| = \frac{1}{2} \sum_k \left| \int_{\overset{m}{g_k}} \frac{(e\xi)\, d\omega}{K(\xi)} \right|,$$

where e is the unit vector in the direction AB. The function $K(\xi)$ is continuous, so that there exists an a such that $K(\xi) < a$. Hence,

$$S^m \geqslant \frac{1}{2a} \sum_k \left| \int_{\overset{m}{g_k}} (e\xi)\, d\omega \right|.$$

Denote by M^m the set of domains $\overset{m}{g_k}$ in which $|\xi e| > \varepsilon > 0$. Then

$$S^m \geqslant \frac{\varepsilon}{2a} \left| \int_{M^m} d\omega \right|.$$

When ε is sufficiently small and m sufficiently large, the diameters of the domains $\overset{m}{g_k}$ are small and

$$\int_{M^m} d\omega > \frac{S}{2},$$

where S is the area of the hypersphere Ω. Thus, when m is sufficiently large, the area of the projection of the polyhedron P^m onto the hyperplane α is

$$S^m > \frac{\varepsilon S}{4a}.$$

Perform a symmetrization of the polyhedron P^m with respect to the hyperplane α. Then, its volume V^m will remain the same, and the section Q^m by the hyperplane α will be of area S^m. The total volume of two cones with vertices A and B and common base Q^m is no larger than V^m. Therefore,

$$V^m \geqslant \frac{1}{n} S^m d_m \geqslant \frac{1}{n} d_m \frac{\varepsilon S}{2a}.$$

We shall show now that the volume V^m is bounded from above. The surface area of the polyhedron is

$$\sigma^m = \sum_k \sigma_k^m \leqslant \sum_k \int_{g_k^m} \frac{d\omega}{K(\xi)}.$$

Since $K(\xi)$ is strictly positive, there exists a b such that $K(\xi) > b > 0$ and

$$\sigma^m \leqslant \frac{1}{b} S.$$

Applying the isoperimetric inequality to the surface area of P^m, we have an estimate for the volume:

$$V^m < V^0.$$

Thus,

$$\frac{1}{n} d_m \frac{\varepsilon S}{2a} < V^0.$$

and we have obtained a uniform estimate for d_m that does not depend on m.

By Blaschke's theorem, the uniformly bounded sequence of polyhedra P^m contains a convergent subsequence. The limit hypersurface P^0 cannot be degenerate. This is true because, if it were, the polyhedron P^m would, for sufficiently large m, be of an arbitrarily small width in a certain direction t^m. Then the projection of this polyhedron onto the hyperplane parallel to t^m would be of arbitrarily small area since the diameter d_m is bounded. But as was shown above, the projection area is bounded from below by $\varepsilon S/2a$. Thus, P^0 cannot be degenerate

Since the surface area functions of the polyhedra P^m converge weakly to the surface area function of the hypersurface P^0, it follows that

$$\sigma(M) = \int_M \frac{d\omega}{K(\xi)} .$$

Hence, the limit hypersurface has Gaussian curvature $K(\xi)$. The theorem is proved.

A. D. Aleksandrov [3] has generalized Minkowski's theorem by proving the existence of a convex hypersurface with an arbitrary surface function. Specifically, let σ denote a nonnegative completely additive function on a unit hypersphere Ω such that

$$\int_\Omega \xi \, d\sigma = 0, \qquad \int_\Omega |e\xi| \, d\sigma > a > 0$$

for any unit vector e. Then there exists a convex hypersurface, unique up to translation, for which σ is a surface area function.

§3. REGULAR SOLUTION OF MINKOWSKI'S PROBLEM

In the previous section, we obtained a generalized solution of Minkowski's problem on the existence of a closed convex hypersurface with given Gaussian curvature $K(\xi)$. The object of this section is to prove the existence of a regular solution of the Minkowski problem when $K(\xi)$ is sufficiently regular. Specifically, we shall prove the following theorem.

Theorem 1. *Let $K(\xi)$ denote a function defined on the unit hypersphere Ω. Suppose that it is positive and regular of class C^m, where $m \geqslant 3$, and that it satisfies the condition*

$$\int_{\Omega} \frac{\xi \, d\omega}{K(\xi)} = 0.$$

*Then there exists a convex hypersurface, regular of class $C^{m+1,\alpha}$ (where $\alpha > 0$) and unique up to translation, whose Gaussian curvature is $K(\xi)$. If the function $K(\xi)$ is analytic, the hypersurface is also analytic.**

1. *A priori* estimates for the radii of normal curvature of a convex hypersurface. Let F denote a regular convex hypersurface in $(n+1)$-dimensional Euclidean space. Let $K(\xi)$ denote the Gaussian curvature of F, given as a function of the unit vector with exterior normal ξ. Let $1/K(\xi) = \varphi(\xi)$. The function φ of the unit vector $\xi \in \Omega$ can be regarded as a function on the spherical image of the hypersurface.

Theorem 2. *The radii of normal curvature of the hypersurface F admit the a priori estimate*

$$R \leqslant \max_{\xi, \gamma} \varphi^{\frac{1}{n}} \left[1 + \frac{1}{n-1}\left(\frac{\varphi'^2}{\varphi^2} - \frac{\varphi''}{\varphi}\right)\right]^{1-\frac{1}{n}},$$

where φ is differentiated at the point ξ with respect to arc length of the great circle on the spherical image of the hypersurface F; the maximum is over all ξ and γ.

PROOF. Let $R(\xi, \gamma)$ denote the radius of normal curvature of the hypersurface F at the point with exterior normal ξ in the direction γ. Let A be a point of F at which the function R attains its maximum with $\xi = \xi_0$ and $\gamma = \gamma_0$. We shall introduce Cartesian coordinates $z, x_1, \ldots, x_n$ in E^{n+1} in the following way. The origin O is a point on the interior normal to F at the point A, the z axis is the straight line OA, and the axes $x_1, x_2, \ldots, x_n$ are parallel to

*See note on page 2.

the principal directions of F at the point A. Suppose that the x_1 axis is directed in the direction in which the radius of normal curvature has its maximum value. By Rodrigues' theorem the x_1 axis has the direction γ_0.

Let H denote the support function of the hypersurface F. Suppose that

$$H_0 = R_1 (z^2 + x_1^2 + \dots + x_n^2)^{1/2}$$

is the support function of a hypersphere of radius R_1. Since the radii of normal curvature of the hypersurface F are no greater than R_1, it follows that $d^2H \leqslant d^2H_0$. In particular, $\partial^2 H/\partial x_1^2 \leqslant \partial^2 H_0/\partial x_1^2$. Let us consider the normalized support functions

$$h(x_1, \dots, x_n) = H(1, x_1, \dots, x_n),$$
$$h_0(x_1, \dots, x_n) = H_0(1, x_1, \dots, x_n).$$

For these functions, the inequality $\partial^2 h/\partial x_1^2 \leqslant \partial^2 h_0/\partial x_1^2$ also holds with equality holding at the point A $(x_1 = x_2 = \dots x_n = 0)$. Let us define

$$w_1 = \left(\frac{\partial^2 h}{\partial x_1^2} - \frac{\partial^2 h_0}{\partial x_1^2}\right) \frac{(1 + x_1^2 + \dots + x_n^2)^{3/2}}{(1 + x_2^2 + \dots + x_n^2)} .$$

The function w_1 attains its maximum (equal to zero) at the point A, and, since

$$\frac{\partial^2 h_0}{\partial x_1^2} \frac{(1 + x_1^2 + \dots + x_n^2)^{3/2}}{1 + x_2^2 + \dots + x_n^2} = R_1,$$

the function

$$w = \frac{\partial^2 h}{\partial x_1^2} \frac{(1 + x_1^2 + \dots + x_n^2)^{3/2}}{1 + x_2^2 + \dots + x_n^2}$$

attains its maximum R_1 at A. Hence, to estimate the maximum of the radii of normal curvature of the hypersurface F is to estimate the maximum of the function w at the point A ($x_1 = x_2 = \ldots = x_n = 0$).

We shall denote by h_i the derivative of h with respect to x_i. Then the function h, which defines the hypersurface F, satisfies the equation

$$(1 + x_1^2 + \ldots + x_n^2)^{\frac{n}{2}+1} \det(h_{ij}) = \varphi \tag{1}$$

(see subsection 2, Sec. 1) and we have

$$w = h_{11} \frac{(1 + x_1^2 + \ldots + x_n^2)^{3/2}}{1 + x_2^2 + \ldots + x_n^2}. \tag{2}$$

At the point A, i.e., when $x_1 = x_2 = \ldots x_n = 0$, we have (due to the special choice of coordinates) $h_{ii} = R_i$, $h_{ij} = 0$ when $i \neq j$, where the R_i are the principal radii of curvature of the hypersurface at the point A. Differentiating (2) at the point A, where w attains its maximum, we obtain

$$w_i = h_{11i} = 0, \tag{3}$$

$$w_{11} = (h_{11})_{11} + 3R_1 \leqslant 0, \quad w_{ii} = (h_{11})_{ii} + R_1, \quad i \neq 1. \tag{4}$$

Differentiating (1) with respect to x_1 at the point A and noting that $h_{ij} = 0$ when $i \neq j$, we get

$$\sum_i h_{11} \ldots (h_{ii})_1 \ldots h_{nn} = \varphi_1,$$

or

$$\varphi \sum_i \frac{(h_{ii})_1}{h_{ii}} = \varphi_1. \tag{5}$$

Twice differentiating (1) with respect to x_1 at the point A, we obtain

$$(n+2)\varphi+\varphi\sum_i \frac{(h_{ii})_{11}}{h_{ii}}+\varphi\sum_{i\neq j}\frac{(h_{ii})_1}{h_{ii}}\frac{(h_{jj})_1}{h_{jj}}-$$
$$-\varphi\sum_{i\neq j}\frac{(h_{ij})_1^2}{h_{ii}h_{jj}}=\varphi_{11}. \quad (6)$$

Taking (5) into account, we conclude that

$$\sum_{i\ j}\frac{(h_{ii})_1}{h_{ii}}\frac{(h_{jj})_1}{h_{jj}}\leqslant\left(\frac{\varphi_1}{\varphi}\right)^2. \quad (7)$$

From (4) we get at the point A

$$\frac{(h_{11})_{11}}{h_{11}}\leqslant-3,\quad \frac{(h_{ii})_{11}}{h_{ii}}\leqslant-\frac{R_1}{R_i}\quad \text{for}\quad i\neq 1. \quad (8)$$

It now follows from Eq. (6) and inequalities (7) and (8) that

$$(n-1)\varphi-\varphi R_1\sum_{i>1}\frac{1}{R_i}+\frac{\varphi_1^2}{\varphi}\geqslant\varphi_{11}. \quad (9)$$

Noting that

$$\sum_{i>1}\frac{1}{R_i}\geqslant(n-1)\left(\frac{1}{R_2}\cdots\frac{1}{R_n}\right)^{\frac{1}{n-1}}=(n-1)\left(\frac{R_1}{\varphi}\right)^{\frac{1}{n-1}},$$

we obtain from inequality (9)

$$(n-1)\varphi-(n-1)\varphi R_1\left(\frac{R_1}{\varphi}\right)^{\frac{1}{n-1}}+\frac{\varphi_1^2}{\varphi}\geqslant\varphi_{11}, \quad (10)$$

so that

$$R_1\leqslant\varphi^{\frac{1}{n}}\left[1+\frac{1}{n-1}\left(\frac{\varphi_1^2}{\varphi^2}-\frac{\varphi_{11}}{\varphi}\right)\right]^{1-\frac{1}{n}}. \quad (11)$$

Let γ_0 be the great circle on the unit hypersphere $z_2 + x_1^2 + \ldots + x_n^2 = 1$ in the direction x_1. Then

$$x_1 = \tan s,$$

where s is the arc of γ_0. Then for $s = 0$, i.e., for $x_1 = 0$,

$$\varphi_1 = \varphi_s', \qquad \varphi_{11} = \varphi_{ss}''.$$

Since R_1 is the maximum radius of normal curvature of the hypersurface F, we obtain the estimate asserted in the theorem:

$$R \leqslant \max_{\xi,\gamma} \varphi^{\frac{1}{n}} \left(1 + \frac{1}{n-1}\left(\frac{\varphi'^2}{\varphi^2} - \frac{\varphi''}{\varphi}\right)\right)^{1-\frac{1}{n}}.$$

Here φ is differentiated with respect to the arc length of the great circle γ through the point ξ. The maximum is over all points ξ of the hypersphere and all directions γ.

2. Auxiliary formulas. Suppose that $z = u(x_1, \ldots, x_n)$ is a convex hypersurface of positive Gaussian curvature. Then the form

$$dz^2 = \frac{\partial^2 u}{\partial x_i \partial x_j} dx_i dx_j$$

is positive-definite. Following E. Calabi [10], we introduce a Riemannian metric defined by the metric tensor

$$g_{\alpha\beta} = \frac{\partial^2 u}{\partial x_\alpha \partial x_\beta}.$$

Let us write

$$\det(g_{\alpha\beta}) = \Phi, \quad \ln\sqrt{\Phi} = \varphi.$$

For the metric in question, the Christoffel symbols of the first kind are

$$\Gamma_{ijk} = \frac{1}{2}\frac{\partial^3 u}{\partial x_i \partial x_j \partial x_k} = A_{ijk}.$$

The quantities A_{ijk} are symmetric with respect to all indices.

Note that

$$g^{\alpha\beta} A_{\alpha\beta k} = \varphi_k. \tag{1}$$

This relation is obtained by differentiating the equation det $(g_{\alpha\beta})$ $= \Phi$. It follows from the expression for the covariant derivative

$$A_{ijk,\,l} = \frac{1}{2} \frac{\partial^4 u}{\partial x_i \partial x_j \partial x_k \partial x_l} + g^{hm} (A_{hjk} A_{mil} + A_{ihk} A_{mjl} + A_{ijh} A_{mkl})$$

that it is symmetric in each pair of indices. In particular,

$$A_{ijk,\,l} = A_{ijl,\,k}. \tag{2}$$

Using well-known formulas, we get the Riemann tensor

$$R_{ijkl} = g^{hm} (A_{hil} A_{mjk} - A_{hik} A_{mjl}). \tag{3}$$

Taking (1) into account and making use of the expression for the Riemann tensor (3), we find the Ricci tensor R_{ik} and the scalar curvature R. Specifically,

$$R_{ik} = g^{jl} R_{ijkl} = g^{jl} g^{hm} (A_{hil} A_{mjk} - A_{hik} A_{mjl}) =$$
$$= g^{jl} g^{hm} A_{hil} A_{mjk} - g^{hm} A_{hik} \varphi_m.$$

Setting

$$\bar{R}_{ik} = g^{jl} g^{hm} A_{hil} A_{mjk},$$

we get

$$R_{ik} = \bar{R}_{ik} - g^{hm} A_{hik} \varphi_m. \tag{4}$$

The scalar curvature is

$$R = g^{ik} R_{ik} = g^{ik} \bar{R}_{ik} - g^{hm} \varphi_h \varphi_m.$$

Setting

$$\bar{R} = g^{ik}\bar{R}_{ik} = A^{ijk}A_{ijk}.$$

we get

$$R = \bar{R} - g^{hm}\varphi_h\varphi_m. \tag{5}$$

Note that

$$\frac{1}{2}\Delta\bar{R} = A^{ijk}\varphi_{i,\,jk} + \bar{R}_{ij}\bar{R}^{ij} + R_{ijkl}R^{ijkl} + A_{ijk,\,\alpha}A^{ijk,\,\alpha}, \tag{6}$$

where Δ denotes the Laplace-Beltrami operator. Actually the proof of this equation is contained in [10]. We shall reproduce it here. First of all, the operator Δ applies to the tensor A_{ijk}. The following exposition rests heavily on the fact that the first covariant derivative $A_{ijk,l}$ of A_{ijk} is symmetric, on Eq. (1), and on the Ricci identity for the difference between the second covariant derivatives. We have

$$\begin{aligned}(\Delta A)_{ijk} = g^{lm}A_{ljk,\,im} = g^{lm}A_{ljk,\,mi} + g^{lm}(A_{ljk,\,im} - A_{ljk,\,mi}) =\\ g^{lm}A_{ljk,\,mi} + g^{lm}(A_{hjk}R^h_{lim} + A_{lhk}R^h_{jim} + A_{ljh}R^h_{kim}) =\\ \varphi_{j,\,ki} + A^{hlm}(A_{hli}A_{mjk} + A_{hlj}A_{mki} + A_{hlk}A_{mij}) -\\ 2A^\alpha_{i\beta}A^\beta_{j\gamma}A^\gamma_{k\alpha}.\end{aligned}$$

Now

$$\begin{aligned}\frac{1}{2}\Delta\bar{R} = \frac{1}{2}g^{lm}\bar{R}_{,\,lm} = A^{ijk}(\Delta A)_{ijk} + A^{ijk,\,l}A_{ijk,\,l} =\\ A^{ijk}\varphi_{i,\,jk} + 3A^{ijk}A^{hlm}A_{hli}A_{mjk} - 2A^{ijk}A^\alpha_{i\beta}A^\beta_{j\gamma}A^\gamma_{k\alpha} +\\ A^{ijk,\,l}A_{ijk,\,l} = A^{ijk}\varphi_{i,\,jk} + \bar{R}_{ij}\bar{R}^{ij} + R_{ijkl}R^{ijkl} + A^{ijk,\,l}A_{ijk,\,l}.\end{aligned}$$

This proves Eq. (6).

Let us define

$$B_{\alpha\beta\gamma ijkl} = \frac{1}{2}(A_{\alpha\beta\gamma,\,i}A_{jkl} - A_{\alpha\beta\gamma}A_{jkl,\,i}).$$

One can easily verify that

$$B_{\alpha\beta\gamma ijkl}B^{\alpha\beta\gamma ijkl} = \frac{1}{2}\bar{R}A_{\alpha\beta\gamma,\,i}A^{\alpha\beta\gamma,\,i} - \frac{1}{8}g^{ij}\bar{R}_{,\,i}\bar{R}_{,\,j}.$$

Using Eqs. (6) and (7), we obtain

$$\Delta\sqrt{\bar{R}} = \frac{\Delta\bar{R}}{2\sqrt{\bar{R}}} - \frac{g^{ij}\bar{R}_{,\,i}\bar{R}_{,\,j}}{4\bar{R}} =$$

$$\frac{1}{\sqrt{\bar{R}}}(\bar{R}_{ij}\bar{R}^{ij} + R_{ijkl}R^{ijkl} + A^{ijk}\varphi_{i,\,jk}) + \frac{2}{\bar{R}^{3/2}}B_{\alpha\beta\gamma ijkl}B^{\alpha\beta\gamma ijkl}.$$

Since the last term in the right-hand side is nonnegative, we have

$$\Delta\sqrt{\bar{R}} \geqslant \frac{1}{\sqrt{\bar{R}}}(\bar{R}_{ij}\bar{R}^{ij} + R_{ijkl}R^{ijkl} + A^{ijk}\varphi_{i,\,jk}). \qquad (8)$$

For brevity, let us write

$$\sqrt{\bar{R}} = \sqrt{A^{\alpha\beta\gamma}A_{\alpha\beta\gamma}} = \psi.$$

Lemma. *The following inequality holds:*

$$\Delta\psi \geqslant \frac{n+1}{n(n-1)}\psi^3 + c_1\psi^2 + c_2\psi + c_3 + c_0\,|\,\partial\psi/\partial x\,|, \qquad (9)$$

where $|\,\partial\psi/\partial x\,|$ *is the maximum of the absolute values of the first derivatives of* φ *with respect to* x_k; c_0, c_1, c_2, *and* c_3 *can be estimated in terms of the second derivatives of* $u(x)$ *and the derivatives of order up to 3 of the function* φ.

PROOF. Since

$$\psi = \sqrt{A^{\alpha\beta\gamma}A_{\alpha\beta\gamma}},$$

one obtains from the expressions for R_{ik} and $\bar{R}_{ik}$

$$R_{ij}R^{ij} = \bar{R}_{ij}\bar{R}^{ij} + a_1\psi^3 + a_2\psi^2, \qquad (10)$$

where a_1 and a_2 can be estimated in terms of the second derivatives of $u(x)$ and the first derivatives of $\varphi(x)$. Furthermore,

$$A^{ijk}\varphi_{i,jk} = b_0\psi \mid \partial\psi/\partial x \mid + b_1\psi^3 + b_2\psi^2 + b_3\psi, \quad (11)$$

where b_0, b_1, b_2, b_3 can be estimated in terms of the second derivatives of $u(x)$ and the third derivatives of $\varphi(x)$. Indeed,

$$\varphi_{i,\,j} = \varphi_{ij} + \varphi_\alpha g^{\alpha s} A_{sij},$$
$$\varphi_{i,\,jk} = \varphi_\alpha g^{\alpha s} A_{sijk} + (*),$$

where (*) denotes a quadratic (in the quantities $A_{\lambda\mu\nu}$) expression with coefficients which admit an estimate. Thus,

$$A^{ijk}\varphi_{i,\,jk} = \varphi_\alpha g^{\alpha s} A^{ijk} A_{ijks} + (**).$$

Obviously, the first term in the right-hand side of the equation is estimated by the quantity $\psi \mid \partial\psi/\partial x \mid + O(\psi^3)$.

In the paper [10], the following inequalities were obtained for a positive-definite Riemannian metric:

$$\frac{1}{n} R^2 \leqslant R_{ij}R^{ij}, \qquad \frac{2}{n-1} R_{ij}R^{ij} \leqslant R_{ijkl}R^{ijkl}.$$

Hence,

$$R_{ij}R^{ij} + R_{ijkl}R^{ijkl} \geqslant \frac{n+1}{n(n-1)} R^2.$$

Noting that $R = \bar{R} - g^{hm}\varphi_h\varphi_m$ and making use of (10), we obtain

$$\bar{R}_{ij}\bar{R}^{ij} + R_{ijkl}R^{ijkl} \geqslant \frac{n+1}{n(n-1)} \psi^4 + d_1\psi^3 + d_2\psi^2. \quad (12)$$

We now obtain inequality (9) from inequality (8) with the aid of Eqs. (11) and (12). This completes the proof of the lemma.

3. An *a priori* estimate for the third derivatives of a support function of a convex hypersurface. Let $z(x_1, \ldots, x_n)$ denote a convex function, regular of class C^5, which satisfies the differential equation

$$\det (z_{ij}) = \Phi > 0, \tag{1}$$

in the domain G. Here (z_{ij}) is the Hessian matrix of the function z and Φ is a positive function, regular of class C^3 and defined on the domain G.

Theorem 3. *At each interior point of the domain G, the third derivatives of the solution of Eq.* (1) *admit an estimate which depends only on the second derivatives of the solution, the derivatives up to third order of the function* Φ, *and the distance from this point to the boundary of G.*

PROOF. Since the function $z(x_1, \ldots, x_n)$ is convex and $\det (z_{ij})$ is positive, the quadratic from d^2x is positive-definite. Introduce in the domain G the Riemannian metric defined by the line element d^2z ($g_{\alpha\beta} = z_{\alpha\beta}$). Let us define

$$\psi = \frac{1}{2} (g^{i\alpha} g^{i\beta} g^{k\gamma} z_{ijk} z_{\alpha\beta\gamma})^{1/2}.$$

According to the lemma in subsection 2, we have the inequality

$$\Delta\psi \geqslant \frac{n+1}{n(n-1)} \psi^3 + c_1\psi^2 + c_2\psi + c_3 + c_0 | \partial\psi/\partial x |, \tag{2}$$

where $| \partial\psi / \partial x |$ is the maximum of the absolute values of the first derivatives of ψ with respect to x_k and the estimates for c_0, c_1, c_2, and c_3 depend on the second derivatives of the function z and the derivatives of Φ up to third order.

Let O be an arbitrary interior point of G, and let ρ be the distance from O to the boundary of the domain G. Denote by ω the ball of radius ρ with center at the point O. Inside the ball ω, we consider the function

$$w = \psi\lambda, \qquad \lambda = \rho^2 - \sum_k (x_k)^2.$$

The function w is nonnegative and it vanishes on the surface of the ball; hence, it attains its maximum w_0 at an interior point A of ω. At the point A,

$$w_\alpha = \partial w/\partial x_\alpha = 0 \quad (\alpha = 1, 2, \ldots, n).$$

Therefore, at that point,

$$\psi_\alpha = w\left(\frac{1}{\lambda}\right)_\alpha, \qquad \psi_{\alpha\beta} = \frac{w_{\alpha\beta}}{\lambda} + w\left(\frac{1}{\lambda}\right)_{\alpha\beta}.$$

Let us substitute the values of the derivatives of ψ into inequality (2). We obtain

$$\Delta\psi = g^{\alpha\beta}\psi_{,\,\alpha\beta} = g^{\alpha\beta}\psi_{\alpha\beta} + g^{ik}\Gamma^j_{ik}\psi_j,$$

where the Γ^j_{ik} are the Christoffel symbols of the second kind for the metric defined by d^2z. Expressions for them were given in subsection 2. What is important for us now is that these expressions contain only the second derivatives of the function z and are linear in third derivatives of Φ. The function w attains its maximum at the point A. Thus, at that point,

$$g^{\alpha\beta}w_{\alpha\beta} \leqslant 0.$$

Therefore,

$$\Delta\psi \leqslant g^{\alpha\beta}w\left(\frac{1}{\lambda}\right)_{\alpha\beta} + c^\alpha w\left(\frac{1}{\lambda}\right)_\alpha, \tag{3}$$

where the c^α are the coefficients, which admit an estimate from above. From inequalities (2) and (3) one obtains an inequality for w at the point A, i.e., an inequality for w_0, namely,

$$\frac{n+1}{n(n-1)} w_0^3 + Q_2(w_0) \leqslant 0, \tag{4}$$

where Q_2 is a second-degree polynomial in w_0 with coefficients which can be estimated in terms of the second derivatives of z and the derivatives of Φ up to third order.

The existence of an estimate for w_0 follows in an obvious manner from inequality (4). Now, since at the center O of the ball

$$\psi \leqslant w_0 / \rho^2,$$

one obtains an estimate for ψ there also. This estimate depends on the above-mentioned derivatives of z and Φ and also on the distance ρ from the point O to the boundary of the domain G.

Since the form d^2z is positive-definite and the second derivatives of z are bounded, the eigenvalues of this form and, hence, the eigenvalues of the form $g^{\alpha\beta}dx_\alpha dx_\beta$ admit an estimate. Therefore, an estimate for

$$\psi^2 = g^{i\alpha} g^{j\beta} g^{k\gamma} z_{\alpha\beta\gamma} z_{ijk}$$

implies an estimate for the third derivatives z_{ijk}, which depends on the same quantities, i.e., on the second derivatives z_{ij}, the derivatives of Φ up to third order, and the distance from the point to the boundary of the domain G. This completes the proof of the theorem.

Theorem 4. *Let F denote a closed convex hypersurface that is regular of class C^5 and that has positive Gaussian curvature $K(\xi)$. Suppose that $H(x_0, x_1, \ldots, x_n)$ is the support function of the hypersurface. Then for the function H and its derivatives up to third order one can get an estimate which depends only on the Gaussian curvature $K(\xi)$ and its derivatives up to third order.*

PROOF. According to Theorem 2 (subsection 1), the radii of normal curvature of the hypersurface F admit an estimate. Since the eigenvalues of the quadratic form d^2H on the unit hypersphere are the principal radii of curvature, the existence of an estimate for the second derivatives H_{ij} on the unit hypersurface is evident.

The existence of an estimate for the function H and its first derivatives is a consequence of the facts that the diameter of F admits an estimate (see subsections 2 and 3, Sec. 2) and that the origin of the coordinate system is inside the hypersurface. Let us now look at any estimate for the third derivatives.

Let ω: $x_0^2 + x_1^2 + \ldots + x_n^2 = 1$ denote a unit hypersphere with center at the origin O of the coordinate system. Fix a certain point A on ω and estimate the third derivatives of the support function H at this point. Introduce new coordinates with the straight line OA as the x_0 axis. Let us set

$$h(x_1, \ldots, x_n) = H(1, x_1, \ldots, x_n).$$

The function h satisfies the differential equation

$$\det(h_{ij}) = \frac{1}{K}(x_1^2 + \ldots + x_n^2 + 1)^{-\frac{n}{2}-1}.$$

The estimate for the second derivatives of H guarantees an estimate for the second derivatives of h in the domain ω': $x_1^2 + \ldots + x_n^2 < 1$. The estimate for the second derivatives of h in ω' guarantees an estimate for the third derivatives of h at the center of the domain ω', i.e., the point A. The third derivatives of the function H at the point A can be expressed in an obvious manner in terms of the derivatives of h and thus can be estimated. This completes the proof of the theorem.

4. The proof of Theorem 1. One of the assertions of Theorem 1 is the existence of a regular convex hypersurface F such that a given function $K(\xi)$ defined on the unit hypersphere is the Gaussian curvature of F. If we put $\varphi(\xi) = 1/K(\xi)$, we have the problem of existence of a solution of the differential equation

$$D(H, \ldots, H) = \varphi(\xi) \tag{1}$$

subject to the condition

$$\int_{\Omega} \xi \varphi(\xi)\, d\omega = 0,$$

where D denotes the sum of principal minors of the Hessian (H_{ij}) of a function H that is positively homogeneous of first degree on the unit hypersphere. Set

$$\varphi(\xi, t) = 1 - t + t\varphi(\xi), \quad 0 \leqslant t \leqslant 1,$$

and consider the equation with parameter t

$$D(H, \ldots, H) = \varphi(\xi, t) \tag{2}$$

subject to the condition

$$\int_{\Omega} \xi\varphi(\xi, t)\, d\omega = 0.$$

To prove that (1) has a solution it is sufficient to prove the following two assertions about Eq. (2).

1. *If (2) has a solution for a certain t_0, then it has a solution for all t sufficiently close to t_0.*

2. *If (2) has a solution for each of a sequence of parameters $t_1, t_2, \ldots, t_n$ converging to t_0, then it has a solution for $t = t_0$.*

It follows from these two assertions that the set of values of the parameter t for which Eq. (2) is solvable is both open and closed. As it is not empty (there exists a trivial solution for $t = 0$), it coincides with the interval $[0,1]$. In particular, Eq. (1) has a solution for $t = 1$. Thus, to prove that Eq. (1) is solvable, it is sufficient to prove assertions 1 and 2.

To prove the first assertion, suppose that H depends on the parameter t. Denote the result of differentiation of $D(H, \ldots, H)$ with respect to t by $D(H, \ldots, Z)$, where $Z = \partial H/\partial t$. Consider now the equation

$$D(H, \ldots, H, Z) = \psi(\xi), \tag{3}$$

where H is a support function of a convex hypersurface with positive Gaussian curvature and Z is an unknown function that is positively homogeneous of first degree. It turns out that Eq. (3) is a self-adjoint elliptic equation with respect to Z [7]. It is known that the problem of solvability of Eq. (3) is related to the problem of nontrivial solutions of the corresponding homogeneous equation

$$D\,(H,\ldots, H, Z) = 0. \tag{4}$$

When such solutions exist, they must be orthogonal to the right-hand side ψ of Eq. (3). So, first of all, we shall show that there exist exactly $n + 1$ independent solutions of Eq. (4), namely,

$$Z_0 = x_0, \quad Z_1 = x_1, \ldots, Z_n = x_n. \tag{5}$$

It follows from Eq. (4) that under a deformation of the hypersurface F with the support function H into the hypersurface F_τ with the support function $H + \tau Z$, the surface area function is stationary for $\tau = 0$. From this, using the same arguments as in the proof of uniqueness in the Minkowski problem in subsection 1, one concludes that the Minkowski inequality for bodies T and T_τ becomes

$$V^n\,(T_\tau, T,\ldots, T) = V^{n-1}(T)\,V\,(T_\tau) + O\,(\tau^2).$$

It follows from the proof of the Minkowski inequality [2] that bodies T and T_τ must be homothetic with accuracy up to τ^2 i.e.,

$$H_\tau = H + \tau Z = \lambda H + \tau \Sigma\, a_k x_k + O\,(\tau^2).$$

But the Gaussian curvature is stationary for $\tau = 0$, so that $\lambda = 1$. Hence, $Z = \Sigma a_k x_k$; i.e., equation (4) has no other solution than (5). Now, the solvability conditions for the nonhomogeneous Eq. (3) are the system of equations

$$\int_\Omega \xi_k \psi\,(\xi)\,d\omega = 0 \qquad (k = 0, 1, \ldots, n), \tag{6}$$

where the ξ_k are the components of the unit vector ξ or, what amounts to the same thing,

$$\int_{\Omega} \xi \psi(\xi)\, d\omega = 0.$$

The solution itself can be represented in terms of the Green's function. Specifically, on the unit hypersphere

$$H(\eta) = \int_{\Omega} G(\xi, \eta)\, \psi(\xi)\, d\omega_{\xi}.$$

We obtain the solution of the nonlinear Eq. (2) by the method of successive approximations. Let the solution with the support function H correspond to $t = t_0$. Set

$$D(H, H, \ldots, Z) = D(H + Z) - D(H) + R(H, Z).$$

Then, for values of the parameter t close to t_0, we have to solve the equation

$$D(H, \ldots, Z) = \Delta t\, (\varphi(\xi) - 1) + R(H, Z).$$

For this, we use the method of successive approximations [6]. The successive approximations can be found from the equation

$$D(H, \ldots, Z_k) = \Delta t\, (\varphi(\xi) - 1) + R(H, Z_{k-1}). \quad (7)$$

Let us show that the solvability condition for Eq. (7) with respect to Z_k holds, i.e., that

$$\int_{\Omega} \xi\, (\Delta t\, (\varphi(\xi) - 1) + R(H, Z_{k-1}))\, d\omega = 0.$$

For the support function $H + \lambda Z$ (where λ is a parameter),

$$\int_{\Omega} \xi D(H + \lambda Z_{k-1})\, d\omega = 0, \quad (8)$$

$$D(H+\lambda Z_{k-1}) = \sum_s \lambda^s A_s(H, Z_{k-1}) = \lambda D(H, Z_{k-1}) + R_\lambda(H, Z_{k-1}).$$

From Eq. (8), which is an identity with respect to λ, it follows that, for all A_s,

$$\int_\Omega \xi A_s(H, Z_{k-1})\, d\omega = 0,$$

and hence,

$$\int_\Omega \xi R(H, Z_{k-1})\, d\omega = 0.$$

The equation

$$\int_\Omega \xi\varphi(\xi)\, d\omega = 0,$$

holds by virtue of the assumptions of the theorem. Thus, equation (7) has a solution at each step of the successive approximations.

Suppose now that the function $\varphi(\xi)$ is analytic. Then the class $C^{2,\alpha}$ solution obtained for Eq. (2) is also analytic [11]. Now let t_k denote a sequence of values of the parameter t converging to t_0 such that Eq. (2) is solvable for each t_k. Let H_k denote the solution corresponding to the value t_k. We shall show that Eq. (2) is solvable for $t = t_0$. It was shown in subsection 3 that, for the solution H of Eq. (2) and its derivatives up to third order, there exists an estimate depending only on the function φ and its derivatives of order up to 3. Hence, the sequence of solutions H_k contains a subsequence which converges uniformly, along with its second derivatives, on the unit hypersphere. The limit function H of the subsequence is of class $C^{2,\alpha}$ with $\alpha \in (0, 1)$ and satisfies Eq. (2) when $t = t_0$. It follows from the theorem on the analyticity of solutions of elliptic equations that the function H is analytic. Thus we have proved both assertions 1 and 2, and this completes the proof of the theorem for the case in which φ is an analytic function.

To extend this result to the case of a three times differentiable function φ, one approximates it in the norm of C^3 with an analytic function φ_a such that

$$\int_{\Omega} \xi\varphi_a(\xi)\, d\omega = 0.$$

We then solve the problem for an analytic function φ_a and shift to the limit solution as $\varphi_a \to \varphi$. The *a priori* estimates obtained in subsection 3 guarantee the convergence of the sequence of analytic solutions to a function of class $C^{2,\alpha}$, which is a solution of the problem for a given three times differentiable function φ. It follows from the theorem on the regularity of solutions of elliptic equations that, if the function φ is of class C^m, where $m \geqslant 3$, then the solution is of class $C^{m+1,\alpha}$ and, if φ is an analytic function, then the solution is analytic.

§4. GENERALIZATION OF MINKOWSKI'S PROBLEM

An elementary symmetric function of the principal radii of curvature

$$S_\nu = \sum_{i_\alpha \neq i_\beta} R_{i_1} \ldots R_{i_\nu}$$

is called a *curvature function of order* ν of a convex hypersurface F in E^{n+1}. When $\nu = n$, the curvature function is the product of the principal radii of curvature. The Minkowski problem considered in Sec. 3 can be stated as an existence problem for a convex hypersurface with preassigned curvature function of order n: $S_n = \varphi_n(\xi)$. A natural generalization of the Minkowski problem is the existence problem for a convex hypersurface with given curvature function of any given order $\nu < n$. It turns out that a necessary condition for solvability of this problem, similarly to the case of the Minkowski problem, is

$$\int_{\Omega} \xi \varphi_\nu (\xi)\, d\omega = 0.$$

However, as A. D. Aleksandrov has found, this condition is not sufficient. In this section, we shall prove the existence of a convex hypersurface with preassigned curvature function of arbitrary order under an additional restriction on the function $\varphi_\nu(\xi)$. The principal method here, as with the Minkowski problem, is the obtaining of *a priori* estimates.

1. An estimate for the principal radii of curvature of a convex hypersurface. Let F denote a regular closed convex hypersurface with positive curvature in $(n + 1)$-dimensional Euclidean space. Let $\varphi_k(\xi)$ denote the value of the curvature function of order k at the point with exterior normal ξ:

$$S_k(R_1, \ldots, R_n) = \varphi_k(\xi).$$

Put

$$\bar{\varphi}_k = (\varphi_k / C_n^k)^{1/k},$$

where C_n^k is the binomial coefficient.

Theorem 1. *For a convex hypersurface F, the radii of normal curvature admit an estimate*

$$R \leqslant \max_{\xi, \gamma} (\bar{\varphi}_k - \bar{\varphi}_k''), \tag{1}$$

where φ is differentiated with respect to the arc length of a great circle γ passing through the point ξ on the unit hypersphere Ω (the spherical image of the hypersurface F) and the maximum is over all points ξ of the hypersurface and all directions γ.

PROOF. Let H denote the support function of F. Then the principal radii of curvature R_i are the roots of the polynomial

$$P(R) = \det (H_{ij} - R\delta_{ij}) = 0,$$

where H_{ij} are the second derivatives of H on the unit hypersphere Ω (subsection 2, Sec. 2). Set

$$h(x_1, \ldots, x_n) = H(1, x_1, \ldots, x_n).$$

Then taking into account the homogeneity of the function H, one can express the derivatives H_{ij} in terms of the derivatives of the function h. The equation $P(R) = 0$ for the principal radii after elementary transformations can be written in the form

$$\begin{vmatrix} \lambda h_{11} - R & \lambda h_{12} & \ldots & x_1 R \\ \lambda h_{21} & \lambda h_{22} - R & \ldots & x_2 R \\ \cdots & \cdots & \cdots & \cdots \\ x_1 R & x_2 R & \ldots & R(1 + x_1^2 + \ldots + x_n^2) \end{vmatrix} = 0 \qquad (2)$$

see subsection 2, Sec. 1), where

$$h_{ij} = \partial^2 h / \partial x_i \partial x_j, \qquad \lambda = (1 + x_1^2 + \ldots + x_n^2)^{1/2}.$$

The coefficients in the polynomial

$$P(-R) = R^{n+1} + S_1 R^n + \ldots + S_n R \qquad (3)$$

are the elementary symmetric functions of the principal radii of curvature.

As was shown in subsection 1, Sec. 3, one can introduce Cartesian coordinates $x_1 \ldots, x_n$ such that the function

$$w = h_{11} \frac{(1 + x_1^2 + \ldots + x_n^2)^{3/2}}{1 + x_2^2 + \ldots + x_n^2}$$

will attain its maximum at the point O: $x_1 = x_2 = \ldots = x_n = 0$, and this maximum is equal to the maximum radius of normal curvature of the hypersurface F. Thus, we have to estimate this maximum. At the point O,

$$\left.\begin{array}{l} h_{ii}=R_i, \quad h_{ij}=0, \quad \text{for} \quad i \neq j, \\ \qquad w_i=h_{11i}=0, \end{array}\right\} \tag{4}$$

$$w_{11}=(h_{11})_{11}+3R_1 \leqslant 0, \ w_{ii}=(h_{11})_{ii}+R_1 \leqslant 0, \ i>1. \tag{5}$$

Let us compute the derivative $S_k' = \partial S_k/\partial x_1$ at the point O, noting that this derivative is a coefficient in the polynomial $P' = \partial P/\partial x_1$. We have, for $i > 1$,

$$P'=\sum_i R(R_1+R)\ldots(R_{i-1}+R)(R_{i+1}+R)\ldots(R_n+R)\,h_{ii}'.$$

Since $h_{11}' = 0$, one can assume the summation to be from $i = 1$ to $i = n$. The coefficient of R^{n-k} in the polynomial P' is

$$S_{k+1}'=\sum_i h_{ii}' S_k^i,$$

where S_k^i is an elementary symmetric function of the variable $R_1 \ldots, R_{i-1}, R_{i+1}, \ldots, R_n$. Noting that $S_1^i = \partial S_{k+i}/\partial R_i$, we obtain

$$S_{k+1}'=\sum_i h_{ii}' \frac{\partial S_{k+1}}{\partial R_i}.$$

Thus

$$S_k' = dS_k \quad \text{when} \quad dR_i = h_{ii}'. \tag{6}$$

Let us now compute the second derivative S_k'' with respect to x_1 at the point O. For that, we differentiate twice the polynomial P with respect to x_1. Omitting the calculations, we give the value of the coefficient S_k'':

$$S_k''=S_{k-1}^i(h_{11})_{ii}+(k+2)S_k-2S_k^1+S_{k-2}^{ij}(h_{ii}'h_{jj}'-h_{ij}'^2),$$

where S_{k-2}^{ij} is an elementary symmetric function of $R_1, R_2, \ldots,$

excluding R_i and R_j. To simplify the writing, we omit the summation sign here and below. By virtue of inequality (5), we obtain after some manipulations

$$S_k'' \leqslant -R_1(n-k+1)S_{k-1} + kS_k + S_{k-2}^{ij}(h_{ii}'h_{jj}' - h_{ij}'^2).$$

Strengthening this inequality by omitting the term $-h_{ij}'^2$, we get

$$S_k'' \leqslant -R_1(n-k+1)S_{k-1} + kS_k + S_{k-2}^{ij}h_{ii}'h_{jj}'. \qquad (7)$$

Note that

$$S_{k-2}^{ij}h_{ii}'h_{jj}' = d^2S_k \quad \text{when} \quad dR_i = h_{ii}',$$

and the function $(S_k)^{1/k}$ is concave with respect to R_i. We have

$$d^2\,(S_k^{1/k}) \leqslant 0.$$

Hence,

$$d^2S_k \leqslant \left(1-\frac{1}{k}\right)\frac{(dS_k)^2}{S_k} = \left(1-\frac{1}{k}\right)\frac{(S_k')^2}{S_k}.$$

Now inequality (7) can be strengthened as follows:

$$S_k'' \leqslant -R_1(n-k+1)S_{k-1} + kS_k + \left(1-\frac{1}{k}\right)\frac{S_k'^2}{S_k}.$$

Hence

$$R_1 \leqslant \frac{kS_k + \left(1-\frac{1}{k}\right)\frac{S_k'^2}{S_k} - S_k''}{(n-k+1)S_{k-1}}. \qquad (8)$$

Noting that

$$\left(\frac{S_{k-1}}{C_n^{k-1}}\right)^{\frac{1}{k-1}} \geqslant \left(\frac{S_k}{C_n^k}\right)^{\frac{1}{k}},$$

we obtain the desired estimate from (8):

$$R_1 \leqslant \bar{\varphi}_k - \bar{\varphi}_k'', \qquad \bar{\varphi} = \left(\frac{S_k}{C_n^k}\right)^{1/k}.$$

Differentiation with respect to x_1 at the point O can be replaced with differentiation with respect to arc length of the great circle issuing from O in the direction of the x_1 axis (subsection 1, Sec. 3). This completes the proof of the theorem.

2. An estimate for the derivatives of a support function of a convex hypersurface. With the estimate for the radii of normal curvature of a convex hypersurface which was obtained in subsection 1, one can easily estimate the support function H and its derivatives up to second order on a unit hypersphere Ω, assuming that the coordinate origin is inside the hypersurface. Specifically, suppose that the radii of normal curvature are less than or equal to R_0. Suppose that A and B are two points which realize the diameter of the hypersurface. Project the hypersurface onto the two-dimensional plane which contains the straight line AB. The projection is a convex domain bounded by a curve γ with the points A and B on it. Since the radius of curvature of the curve γ is no greater than R_0, the length of this curve is no greater than $2\pi R_0$. Hence, the distance between the points A and B is no greater than πR_0. Since the coordinate origin is inside the hypersurface, the values of H on the unit hypersphere are no greater than πR_0 because they are the distances from the coordinate origin to the support hyperplanes. The derivatives $\partial H/\partial x_i$ are also no greater than πR_0 since they are the coordinates of points on the hypersurface. The second derivatives $\partial^2 H/\partial x_i \partial x_j$ have already been estimated in terms of the radii of normal curvature (see Sec. 3). Thus, for a hypersurface with given curvature function $\varphi_k(\xi)$, the support function on the unit hypersphere and its first and second derivatives can be estimated in terms of $\varphi_k(\xi)$ and its first and second derivatives. Now we shall show that there exist *a priori* estimates for the Hölder constants of the second derivatives.

The principal radii of curvature of the hypersurface satisfy the equation

$$\det\,(H_{ij} - R\delta_{ij}) = 0.$$

Replacing the function H with

$$h(x_1 \ldots, x_n) = H(1, x_1, \ldots, x_n),$$

we obtain

$$\begin{vmatrix} \lambda h_{11} - R & \lambda h_{12} & \ldots & x_1 R \\ \lambda h_{21} & \lambda h_{22} - R & \ldots & x_2 R \\ \ldots & \ldots & \ldots & \ldots \\ x_1 R & x_2 R & \ldots & \lambda^2 R \end{vmatrix} = 0,$$

where $\lambda = (1 + x_1^2 + \ldots + x_n^2)^{1/2}$. It follows from this equation that, with an order of magnitude $O(x^2)$, the principal radii of curvature are the eigenvalues of the matrix (h_{ij}) in the neighborhood of the point $O\,(x = 0)$.

Obviously, a convex hypersurface with given curvature function $\varphi(\xi)$ satisfies an equation of the form

$$\sum_{\alpha,\,\beta,\,\gamma} \begin{vmatrix} h_{\alpha\alpha} & h_{\alpha\beta} & \ldots & h_{\alpha\gamma} \\ h_{\beta\alpha} & h_{\beta\beta} & \ldots & h_{\beta\gamma} \\ \ldots & \ldots & \ldots & \ldots \\ h_{\gamma\alpha} & h_{\gamma\beta} & \ldots & h_{\gamma\gamma} \end{vmatrix} = \Phi, \qquad (1)$$

where the left-hand side is the sum of the kth order principal minors of the matrix $(h_{\alpha\beta})$. The right-hand side is an algebraic expression with respect to the second derivatives $h_{\alpha\beta}$, the given function $\varphi_k\,(\xi)$, and the squared independent variables $x_1^2, x_2^2, \ldots$. At the point O, we have $\Phi = \varphi_k\,(\xi)$ and $\partial\Phi/\partial x_\alpha = \partial\varphi_k/\partial x_\alpha$.

Differentiate Eq. (1) with respect to x_α and define $\partial h/\partial x_\alpha = v$. Then for v we obtain a linear differential equation of the form

$$\sum_{i,\,j} a_{ij} v_{ij} = f. \qquad (2)$$

Consider the eigenvalues of the matrix (a_{ij}). If we rotate the Cartesian coordinates in such a way that the directions of the coordinate axes become the principal directions at the point O, then the coefficients a_{ij} vanish for $i \neq j$ and the coefficients a_{ii} are elementary symmetric functions, of order $k - 1$, of the principal radii of curvature R_β, where $\beta \neq i$. Let O' denote a point close to O. Suppose that we perform a transformation of coordinate system as above, at the point O'. Then the coefficients a_{ij} will differ from those above by quantities of order of magnitude $O(x^2)$. Hence, in a sufficiently small neighborhood of the point O, the eigenvalues of the matrix (a_{ij}) can be represented as

$$\lambda_i = \partial S_k / \partial R_i + O(x^2).$$

It has been shown in [12] that one can obtain an *a priori* estimate, depending on v and the coefficients of the equation, for the Hölder constants of the first derivatives of v if the eigenvalues λ_i of the matrix (a_{ij}) satisfy the condition

$$(n-1)\sum_{i<k} (\lambda_i - \lambda_k)^2 < \left(\sum \lambda_i\right)^2,$$

or

$$(n-1)\sum_i \lambda_i^2 < \left(\sum \lambda_i\right)^2. \tag{3}$$

This means that if this condition is satisfied, then there exists an *a priori* estimate for the Hölder constants of the second derivatives h_{ij} with an exponent $\alpha > 0$, and this estimate depends on the second derivatives h_{ij} and the first derivatives of the curvature function.

Let us show that condition (3) holds whenever the curvature function $\varphi_k(\xi)$ satisfies the inequality

$$\left(1 - \frac{1}{n}\right)^{\frac{1}{2k-1}} \max_{\xi,\, \gamma} (\bar{\varphi}_k - \bar{\varphi}_k'') < \bar{\varphi}_k, \qquad \bar{\varphi}_k = \left(\frac{\varphi_k}{C_n^k}\right)^{\frac{1}{k}}. \tag{4}$$

At the point O,

$$\lambda_i = a_{ii} = S_{k-1}^i (R_1, \ldots, R_{i-1}, R_{i+1}, \ldots, R_n),$$
$$\sum \lambda_i = (n-k+1) S_{k-1} (R_1, \ldots, R_n).$$

Noting that

$$\left(\frac{S_{k-1}}{C_n^{k-1}}\right)^{\frac{1}{k-1}} \geqslant \left(\frac{S_k}{C_n^k}\right)^{\frac{1}{k}} = \bar{\varphi}_k,$$

we obtain

$$\sum \lambda_i \geqslant (n-k+1) C_n^{k-1} \bar{\varphi}^{k-1}. \tag{5}$$

Suppose that R_0 is the maximum of the radii of normal curvature. Then,

$$\sqrt{\sum \lambda_i^2} \leqslant \sqrt{n} C_{n-1}^k R_0^{k-1}. \tag{6}$$

It follows from (5) and (6) that condition (3) is satisfied when inequality (4) is. Since (4) is a strict inequality, condition (3) is satisfied not only at the point O but also in some δ-neighborhood of O, and there exists a positive lower bound for δ.

3. Existence of a convex hypersurface with a preassigned curvature function. Let F denote a closed convex hypersurface and let D_k denote a curvature function of order k. Let us show that

$$\int_{\Omega} \xi D_k d\omega = 0. \tag{1}$$

Construct a convex hypersurface F_t parallel to F at a distance t. By Rodrigues' formula, we have for the hypersurface F

$$dr = Rdn$$

and for the hypersurface F_t

$$d(r + tn) = (R + t)\, dn.$$

This means that, at a point with the same exterior normal n, the hypersurface F_t has the principal radii of curvature $R_i + t$. The curvature function of order n for the hypersurface F_t is

$$D_{nt} = (R_1 + t)\ldots(R_n + t) = D_n + tD_{n-1} + \ldots,$$

where the D_k are the curvature functions of different orders for the hypersurface F. Since the equation

$$\int_\Omega \xi D_{nt} d\omega = 0$$

is an identity in t, the curvature functions of the hypersurface F satisfy

$$\int_\Omega \xi D_k d\omega = 0 \qquad (k = 1, 2, \ldots, n).$$

Thus, condition (1) is necessary for existence of a convex hypersurface with a preassigned curvature function. However, A. D. Aleksandrov has shown that, if $k < n$, condition (1) is not sufficient. We shall prove the existence of a convex hypersurface with a preassigned curvature function under the additional conditions that was used in subsection 2 to obtain *a priori* estimates for the Hölder constants of the second derivatives of the support function.

Theorem 2. *Let $\Phi_k(\xi)$ denote a regular positive function defined on the unit hypersphere Ω and satisfying the conditions*

$$\int_\Omega \xi \Phi_k(\xi)\, d\omega = 0, \quad k > 1, \tag{2}$$

$$\left(1 - \frac{1}{n}\right)^{\frac{1}{2(k-1)}} \max_{\xi,\, \gamma} (\varphi_{kt} - \varphi''_{kt}) < \varphi_{kt}(\xi), \tag{3}$$

where $\varphi_{kt} = (\Phi_{kt}/C_n^k)^{1/k}$, $\Phi_{kt} = t\Phi + 1 - t$, $0 \leqslant t \leqslant 1$. *Then there exists a closed convex hypersurface whose curvature function of order* k *is* $\Phi_k(\xi)$. *When the function* Φ_k *is of class* C^m *(where* $m \geqslant 3$*), the hypersurface is of class* $C^{m+1,\alpha}$, *where* $\alpha > 0$. *When the function* Φ_k *is analytic, the hypersurface is analytic.*

The proof of this theorem is based on reasoning similar to that used in the case of the Minkowski problem (Sec. 3). Specifically, the function $\Phi_k(\xi)$ is included in a family of functions $\Phi_{kt}(\xi)$. Then we have to prove two assertions: 1) if the existence problem is solvable for a particular value of the parameter t, it is also solvable for t sufficiently close to that value; 2) if the problem is solvable for a sequence of values of the parameter t that converges to t_0, it is also solvable for $t = t_0$. We shall not give the complete proof of assertions 1) and 2), but shall indicate only the essential steps of it.

Let $D_k(H)$ denote the curvature function of the hypersurface represented as the sum of the principal minors of the matrix $(H_{\alpha\beta})$. Then the problem of existence of a hypersurface F with a preassigned curvature function Φ_{kt} reduces to the problem of solvability of the differential equation

$$D_k(H) = \Phi_{kt} \tag{4}$$

on the hypersphere Ω. To solve this equation by the method of continuation, we have to study first the following linear equation for the function $Z = \partial H/\partial t$, which is obtained by differentiating Eq. (4) with respect to the parameter t:

$$D_k(H, Z) = \Phi_k - 1. \tag{5}$$

This is a self-adjoint linear (in Z) elliptic differential equation. The operator D_k is self-adjoint for any k since it is a coefficient in the expansion in λ of the self-adjoint operator $D_n(H + \lambda\Omega, Z)$ for a hypersurface whose support function is $H + \lambda\Omega$ and which is parallel to F. It is also elliptic since the hypersurface F_t is of strictly positive normal curvature. One can show that, as with the Minkowski problem, the homogeneous equation corresponding to Eq. (5)

$$D_k\,(H, Z) = 0 \tag{6}$$

has exactly $n+1$ independent solutions:

$$Z_0 = x_0\,, Z_1 = x_1\,, \ldots, Z_n = x_n. \tag{7}$$

The proof is based on the fact that, when a hypersurface with support function H, is subject to a deformation into a hypersurface with support function $H + tZ$, the condition $D_k\,(H, Z) = 0$ expresses the stationarity of the curvature function for $t = 0$. From this, we use Aleksandrov's inequality (subsection 4, Sec. 1) applied to the mixed curvature functions to deduce that, for $t = 0$, this deformation is an infinitesimal translation, i.e., Z is a linear function. Now, since the solutions (7) represent all independent solutions of the homogeneous equation (6), the nonhomogeneous equation (5) is always solvable because the orthogonality conditions

$$\int_{\Omega} \xi\,(\Phi_k - 1)\,d\omega = 0$$

are satisfied.

Since Eqs. (5) are solvable, one can solve Eqs. (4) for t close to t_0 by the method of successive approximations as with Minkowski's problem (Sec. 3). The proof of the second assertion is based on the fact that one can obtain *a priori* estimates for the support function of the supposed solution.

Finally, we note that, to apply the continuation method, we need for the hypersurface in question to be strictly convex. However, if $k < n$, positiveness of the curvature function Φ_{kt} does not, in general, guarantee positiveness of the principal radii of curvature. Let us show that this is guaranteed by inequality (3). If at least one R_i is equal to zero, then

$$\Phi_{kt} < C_{n-1}^k R_0^k,$$

where R_0 is the maximum radius of curvature of the hypersurface. Hence,

$$\varphi_{kt} < \left(\frac{n-k}{n}\right)^{1/k} R_0 \leqslant \left(\frac{n-k}{n}\right)^{1/k} \max_{\xi,\gamma} (\varphi_{kt} - \varphi_{kt}'').$$

But this contradicts condition (3) since

$$\left(\frac{n-k}{n}\right)^{1/k} < \left(1 - \frac{1}{n}\right)^{1/(2k-2)}$$

when $k > 1$.

4. A general estimate for the radii of normal curvature. The result regarding the *a priori* estimate for the radii of normal curvature obtained in subsection 1 can be considerably generalized:

Theorem 3. *Let the function f be defined for $R_1, R_2 \ldots, R_n > 0$ and symmetric with respect to all the variables R_i. Let f be nonnegative and concave, i.e.,*

$$\frac{\partial^2 f}{\partial R_i \partial R_j} \lambda_i \lambda_j \leqslant 0.$$

Suppose that

$$\lim_{R\to\infty} \frac{1}{R} f(R, R, \ldots, R) = \alpha > 0.$$

Then for the principal radii of curvature of a closed convex hypersurface F satisfying the equation

$$f(R_1, R_2, \ldots, R_n) = \varphi(\xi), \tag{1}$$

we have the estimate

$$R \leqslant \max_{\xi, \gamma} \frac{1}{\alpha} (\varphi - \varphi'').$$

Theorem 1 is a consequence of Theorem 3. We set $f \equiv (S_k/C_n^k)^{1/k}$, where S_k is an elementary symmetric function of the principal radii of curvature. One can easily see that this function satisfies the hypotheses of the theorem. For this function, $\alpha = 1$. Therefore, the estimate given by Theorem 3 coincides with the estimate in Theorem 1. Let us first prove Theorem 3 under the additional assumption that the function f is a function of the elementary symmetric function S_k. Later on, this restriction will be removed.

Suppose that the radius of normal curvature of the hypersurface F attains its maximum at a certain point P. Introduce Cartesian coordinates $x_0, x_1, \ldots, x_n$ such that the hyperplane $x_0 = 1$ is tangent to the hypersurface at the point P and the directions of the axes $x_1, \ldots, x_n$ coincide with the principal directions. Suppose that the direction of x_1 corresponds to the maximum radius of curvature R_1.

Let $H(x_0, x_1, \ldots, x_n)$ denote the support function of the hypersurface F. Set

$$h(x_1, \ldots, x_n) = H(1, x_1, \ldots, x_n).$$

Then, as was shown in subsection 1, Sec. 3, the function

$$w = h_{11} \frac{(1 + x_1^2 + \ldots + x_n^2)^{3/2}}{1 + x_2^2 + \ldots + x_n^2} \tag{2}$$

attains its maximum at the point $O(x_1 = x_2 = \ldots = x_n = 0)$. This maximum is the maximum radius of normal curvature of the hypersurface. Let us estimate it.

At the point O,

$$\left.\begin{aligned}&h_{ii}=R_i,\quad h_{ij}=0\quad \text{for}\quad i\neq j,\\&w_i=(h_{11})_i=0,\\&w_{11}=(h_{11})_{11}+3R_1\leqslant 0,\ w_{ii}=(h_{11})_{ii}+R_1\leqslant 0,\quad i>1.\end{aligned}\right\}\tag{3}$$

The quantities S_k on the hypersurface are functions of the variable $x_1, x_2, \ldots, x_n$. We denote differentiation with respect to x_1 by primes. In subsection 1, we obtained the following expressions for the derivatives S_k' and S_k'':

$$S_k' = \sum_i \frac{\partial S_k}{\partial R_i} h_{ii}', \tag{4}$$

$$S_k'' = \sum_i \frac{\partial S_k}{\partial R_i}(h_{11})_{ii} + (k+1)\,S_k - 2\frac{\partial S_{k+1}}{\partial R_1} + \sum_{i,j}\frac{\partial^2 S_k}{\partial R_i\,\partial R_j}(h_{ii}'h_{jj}' - h_{ij}'^2). \tag{5}$$

Consider the second derivative of f with respect to x_1. We have

$$f'' = \sum_k \frac{\partial f}{\partial S_k} S_k'' + \sum_{k,l}\frac{\partial^2 f}{\partial S_k\,\partial S_l} S_k' S_l'.$$

Substituting into this formula the expressions for S_k' and S_l' given by Eqs. (4) and (5) and the expression for $(h_{11})_{ii}$ given by (3), we get:

$$f'' = \sum_i \frac{\partial f}{\partial R_i} w_{ii} - \sum_i (R_1 - R_i)\frac{\partial f}{\partial R_i} + \sum_{i,j}\frac{\partial^2 f}{\partial R_i\,\partial R_j} h_{ii}'h_{jj}' - \sum_{i,j,k}\frac{\partial f}{\partial S_k}\frac{\partial^2 S_k}{\partial R_i\,\partial R_j} h_{ij}'^2. \tag{6}$$

Since the function f is nonnegative and concave, $\partial f/\partial R_i \geqslant 0$. Hence, the first term on the right-hand side of formula (6) is nonpositive. Because of the concavity of the function f, the third term is also nonpositive. Let us show that the last term is nonpositive also. We have

$$\frac{\partial f}{\partial R_i} = \sum_k \frac{\partial f}{\partial S_k} \frac{\partial S_k}{\partial R_i} = R_j \sum_k \frac{\partial f}{\partial S_k} \frac{\partial^2 S_k}{\partial R_i \partial R_j} + \sum_k \frac{\partial f}{\partial S_k} \frac{\partial^2 S_{k+1}}{\partial R_i \partial R_j},$$

$$\frac{\partial f}{\partial R_j} = \sum_k \frac{\partial f}{\partial S_k} \frac{\partial S_k}{\partial R_j} = R_i \sum_k \frac{\partial f}{\partial S_k} \frac{\partial^2 S_k}{\partial R_i \partial R_j} + \sum_k \frac{\partial f}{\partial S_k} \frac{\partial^2 S_{k+1}}{\partial R_i \partial R_j}.$$

Hence,

$$\sum_k \frac{\partial f}{\partial S_k} \frac{\partial^2 S_k}{\partial R_i \partial R_j} = \left(\frac{\partial f}{\partial R_j} - \frac{\partial f}{\partial R_i}\right)(R_i - R_j)^{-1}. \tag{7}$$

Taking into consideration the fact that f is concave and symmetric in the variables R_i and R_j, we conclude from this formula that the last term in the right-hand member of formula (6) is nonpositive.

Differentiation of f with respect to x_1 at the point P can be replaced by differentiation of φ with respect to arc length along the great circle in the direction x_1. Thus,

$$\varphi'' \leqslant - \sum_i (R_1 - R_i) \frac{\partial f}{\partial R_i}. \tag{8}$$

Since f is nonnegative and concave,

$$\sum_k R_k \frac{\partial f}{\partial R_k} \leqslant f, \quad \sum_k \frac{\partial f}{\partial R_k} \geqslant \alpha.$$

Inequality (8) can be refined as follows:

$$\varphi'' \leqslant - \alpha R_1 + \varphi.$$

Hence $R_1 \leqslant (\varphi - \varphi'')/\alpha$.

At the beginning of the proof, it was assumed that f is a function of elementary symmetric functions in the variables R_i.

We shall now remove this restriction. For this we approximate the function f with an analytic function $\bar{f}$ which possesses the properties of f (symmetry, nonnegativeness, concavity); and we approximate the hypersurface F by an analytic hypersurface $\bar{F}$. From $\bar{F}$ and $\bar{f}$ a function $\bar{\varphi}$ can be found: $\bar{\varphi} = \bar{f}(\bar{F})$. Now we apply the estimate obtained and use a limit process $\bar{f} \to f$, $\bar{F} \to F$ in the metric of C^4.

Theorem 4. *Let f denote a symmetric function of the variables $R_1, R_2, \ldots, R_n$ defined for $R_i > 0$. Suppose that it is non-decreasing with respect to each variable. Suppose that the conditions*

$$\partial f/\partial R_i \geqslant c' > 0, \qquad d^2 f/(df)^2 \leqslant c'' < \infty.$$

are satisfied in the domain defined by the inequalities $\min \varphi \leqslant f \leqslant \max \varphi$. *Then, for the principal radii of curvature of a closed convex hypersurface F satisfying Eq.* (1), *there exists an a priori estimate from above.*

Theorem 4 is a consequence of Theorem 3. We need only note that the function

$$f_1 = 1 - e^{-c'' f}$$

satisfies the hypotheses of Theorem 3.

§5. A MULTIDIMENSIONAL ANALOG OF THE MONGE–AMPÈRE EQUATION

In case of two variables x and y, the Monge-Ampère equation is an equation of the form

$$\frac{\partial^2 u}{\partial x^2}\frac{\partial^2 u}{\partial y^2} - \left(\frac{\partial^2 u}{\partial x\, \partial y}\right)^2 = \varphi(x, y).$$

The natural generalization of this equation to the case of n variables is the equation

$$\det (u_{ij}) = \varphi (x_1, \ldots, x_n), \tag{1}$$

where the u_{ij} are the second derivatives of an unknown function $u(x_1, \ldots, x_n)$. This equation will be the subject of study in this section. We always assume the function φ to be positive and the solution of Eq. (1) to be convex. We shall define the concept of a generalized solution and prove the solvability of the Dirichlet problem for this equation in a strictly convex domain. We shall also prove that a strictly convex solution is regular when the function φ is sufficiently regular. Finally, we shall prove that the Dirichlet problem in a strictly convex domain always has a regular solution when the boundary of the domain and the restriction of the unknown function to the boundary are of class C^2.

1. Existence of a convex polyhedron with given areas of the normal images of the vertices. Let $z(x_1, \ldots, x_n)$ denote a convex function and let F denote the convex hypersurface defined by the equation

$$z = z (x_1 \ldots, x_n).$$

Let M denote a point set on the hypersurface F. From it, we define a point set M^* on the hyperplane $z = 0$ as follows: A point P with coordinates $p_1, p_2, \ldots, p_n$ belongs to the set M^* if there exist a point (x, z) in M and a support hyperplane through (x, z) defined by the equation

$$z = p_1 x_1 + \ldots + p_n x_n + c.$$

The set M^* is called the *normal image* of M and the mapping that assigns M to M^* is called the *normal* mapping. When F is a regular hypersurface, the function z is also regular and the normal mapping is given by the formulas

$$p_1 = \partial z/\partial x_1, \quad p_2 = \partial z/\partial x_2, \ \ldots, \quad p_n = \partial z/\partial x_n.$$

The normal mapping possesses many properties of the spherical mapping and can be obtained from the latter by projection from the center of the hypersphere $z^2 + x_1^2 + \ldots + x_n^2 = 1$ onto the hyperplane $z = 1$ and then by parallel projection onto the coordinate hyperplane $z = 0$. Let us note that the normal image of a Borel set is also a Borel set.

Lemma 1. *Let P denote a convex polyhedron in the hyperplane $z = 0$, let $B_1, B_2, \ldots$ denote its vertices, and let $A_1, A_2, \ldots$ denote certain marked points inside the polyhedron. Then there exists a convex function $z(x_1, \ldots, x_n)$ defined on P that is linear on each face of the polyhedron P and assumes given values ψ_h at the vertices B_k of the polyhedron P. The hypersurface $\widetilde{P}$ defined by this function is polyhedral, its vertices are projections of the points A_k, and the normal images of these vertices are of preassigned areas s_k.*

PROOF. Denote by $\widetilde{B}_k$ a point with projection B_k and z-coordinate ψ_k. Construct the convex hull of the points $\widetilde{B}_k$. It is a convex polyhedron (which may be in a hyperplane if the points $\widetilde{B}_k$ lie in a single hyperplane). Denote by $\widetilde{P}$ that part of the surface of the polyhedron that is convex in the direction $z < 0$. Let $\widetilde{z}\,(x_1, \ldots, x_n)$ denote a convex function which defines the hypersurface $\widetilde{P}$. On $\widetilde{P}$ we mark the points $\widetilde{A}_k$ whose projections are the A_k.

Displace the point $\widetilde{A}_1$ in the direction $z < 0$ and construct the convex hull of the points $\widetilde{B}_k$ and $\widetilde{A}_1$. As a result, we obtain a convex polyhedron, and again we consider that part $\widetilde{P}$ of its surface that is convex in the direction $z < 0$. The polyhedron $\widetilde{P}$ has a unique interior vertex, which is $\widetilde{A}_1$. The displacement of $\widetilde{A}_1$ is selected so that the area of its normal image becomes equal to s_1. Next, we displace the point $\widetilde{A}_2$ in such a way that the area of the normal image of $\widetilde{A}_2$ becomes equal to s_2. After this displacement, the area of the normal image of $\widetilde{A}_1$ may change but it will not increase. We now displace $\widetilde{A}_3$ so that the area of its normal image becomes equal to s_3. Again, the areas of the normal images of $\widetilde{A}_1$ and $\widetilde{A}_2$ can only decrease. We repeat this process with respect to all the other vertices and then repeat it over again starting with $\widetilde{A}_1$.

Under the same deformation of the polyhedron $\widetilde{P}$, the function $z(x)$ varies monotonically (nonincreasing), and the areas of the normal images of the vertices $\widetilde{A}_k$ remain less than or equal to s_k. Moreover, the function $z(x)$ remains bounded throughout the deformation. This is true because, if it were unbounded, the normal image of $\widetilde{P}$ would cover any given domain, but this is impossible since the area of its normal image does not exceed Σs_k.

Because of these two properties of the function $z(x)$ (monotonicity and boundedness), the sequence of polyhedra corresponding to the steps of the deformation is convergent. Let us show that, for the limit polyhedron, the areas of the normal images of the vertices $\widetilde{A}_k$ are equal to given numbers s_k. To see this, suppose that, at some vertex $\widetilde{A}_k$ of the limit polyhedron, the area of the normal image is less than s_k. (It is obvious that it cannot be greater than s_k.) Then the same inequality holds for all polyhedra beginning with some step of the deformation. But this is impossible since the displacement of the vertex $\widetilde{A}_k$ at each step of the deformation is selected so that the area of the normal image of $\widetilde{A}_k$ becomes equal to s_k. Thus, the vertices $\widetilde{A}_k$ of the limit polyhedron have areas of normal images equal to s_k.

It is obvious that, throughout the deformation process, the function $z(x)$ remains linear on each face of the polyhedron P and $z(B_k) = \psi_k$. This property also holds for the limit polyhedron. Hence, the limit function $z(x)$, which defines the limit polyhedron, is linear on each face of the polyhedron P and is equal to ψ_k at its vertices B_k. This completes the proof of the lemma.

2. Concept of a generalized solution. The Dirichlet problem. Let $z(x_1, \ldots, x_n)$ denote a convex solution of class C^2 of the equation

$$\det (Z_{ij}) = \varphi \tag{1}$$

with positive right-hand side. Let F denote a hypersurface defined by the equation $z = z(x_1, \ldots, x_n)$, let M denote an arbitrary Borel set on the hypersurface F, let $\overline{M}$ denote its projection onto the hyperplane $z = 0$, and let M^* denote the normal image of the set M. We have

$$\int_{\overline{M}} \det (z_{ij})\, dx_1 \ldots dx_n = \int_{\overline{M}} \varphi\, dx_1 \ldots dx_n.$$

With the aid of the normal mapping, the left-hand side of this equation can be written

$$\int_{\overline{M}} \det (z_{ij})\, dx_1 \ldots dx_n = \int_{M^*} dp_1 \ldots dp_n = S(M^*),$$

where $S(M^*)$ is the area of the normal image of the set M. Thus, if $z(x_1, \ldots, x_n)$ is a convex solution of class C^2 of Eq. (1), then, for any Borel set M,

$$S(M^*) = \int_{\overline{M}} \varphi\, dx_1 \ldots dx_n. \tag{2}$$

Conversely, if for any Borel set M on a hypersurface of class C^2 the normal mapping is such that Eq. (2) holds, then the function $z(x_1, \ldots, x_n)$ defining the hypersurface F is a solution of Eq. (1). But Eq. (2) is meaningful for any convex function $z(x_1, \ldots, x_n)$ without the assumption that it is twice differentiable.* Therefore, we arrive at a natural generalization of the concept of solution of Eq. (1). Specifically, a convex function $z(x_1, \ldots, x_n)$ is called a *generalized solution of Eq.* (1) if it satisfies Eq. (2).

Theorem 1. *Let G denote a strictly convex domain, φ a positive continuous function defined in G, and ψ a continuous function defined on the boundary of G. Then there exists a unique generalized solution of the Dirichlet problem for Eq.* (1).

PROOF. Let us partition the domain G into small domains g_k and mark a point A_k inside each of them. Let us define

$$s_k = \int_{g_k} \varphi\, dx_1 \ldots dx_n.$$

We take a sufficiently dense net of points B_k on the boundary of the domain G and construct the convex hull of it. It is a convex polyhedron P whose vertices are B_k. When the net of the points B_k is sufficiently dense, the points A_k are contained strictly inside the polyhedron P.

*Editor's note. A convex function is twice differentiable almost everywhere.

By Lemma 1 there exists a polyhedron $\widetilde{P}$ the projections of whose interior vertices onto the hyperplane $z = 0$ are the points A_k and the areas of whose normal images are equal to s_k. The function $z(x_1, \ldots, x_n)$ which defines the polyhedron $\widetilde{P}$ is linear on the faces of the polyhedron P and equal to $\psi(B_k)$ at its vertices. Construct a sequence of polyhedra $\widetilde{P}$ by decreasing the diameters of the domains g_k and increasing the density of the net of points B_k. As was shown in subsection 1, the polyhedra $\widetilde{P}$ are uniformly bounded since for each polyhedron the total area of the normal images of the vertices $\bar{A}_k$ is no greater than $\int \varphi dx_1 \ldots dx_n$. Hence, the sequence of the polyhedra $\widetilde{P}$ contain a subsequence which converges to a convex hypersurface $\bar{F}$:

$$z = z(x_1, \ldots, x_n).$$

We assert that the function $z(x_1, \ldots, x_n)$ is the solution of the Dirichlet problem for Eq. (1).

We define two set functions in the domain G: $\widetilde{S}$ and Φ. Suppose that the function $\widetilde{S}$ is zero everywhere except at the points A_k, where it assumes the values s_k. The function Φ is defined by the equation

$$\Phi(M) = \int_M \varphi \, dx_1 \ldots dx_n.$$

Since the function φ depends continuously on the polyhedra $\widetilde{P}$, the corresponding set functions $\widetilde{s}_k$ converge weakly to the function Φ when the sequence $\widetilde{P}$ converges to the hypersurface F. For each polyhedron of $\widetilde{P}$, the set function $\widetilde{S}$ is the area of the normal image of this polyhedron. Since the normal mapping possesses properties similar to those of a spherical mapping (see subsection 1), the areas of the normal images converge weakly to the area of the normal image of the hypersurface F. Hence, the area of the normal image of F is

$$S(M^*) = \int_{\bar{M}} \varphi \, dx_1 \ldots dx_n.$$

This means that function $z(x_1, \ldots, x_n)$ determining the hypersurface F is the generalized solution of Eq. (1).

Let us show that the solution obtained satisfies the boundary condition $z = \psi$. To do this, we need only to show that the limit hypersurface F is projected in a unique manner onto the closed domain $\bar{G}$. It is obvious that nonuniqueness may occur only on the boundary of the domain G. Let A_0 be a point of the hypersurface F that is projected onto the boundary of G. For this point, we have $z < \psi$ (the opposite inequality $z > \psi$ is impossible because of the convexity of the polyhedra $\tilde{P}$). Construct the convex hull of the edge of the hypersurface F. Since the domain G is strictly convex, the point A_0 is outside the convex hull. Hence, there exists a hyperplane α:

$$z = \sum_i a_i x_i + c$$

separating the point A_0 from the edge of the hypersurface F. Let γ denote the intersection of F and the hyperplane α and let V_0 denote a cone which projects γ from the point A_0. The normal image of the cone V_0 is contained in the normal image of the hypersurface F since for each support hyperplane V_0 there exists a parallel support hyperplane of F. The area of the normal image of the cone V_0 is infinite since the normal image of the cone V_0 is an unbounded convex domain. Let $\tilde{P}$ denote a polyhedron sufficiently close to F, let $\tilde{\gamma}$ denote the intersection of $\tilde{P}$ with the hyperplane α, let $\tilde{A}$ denote its closest vertex to A_0, and let $\tilde{V}$ denote a cone projecting $\tilde{\gamma}$ from the vertex $\tilde{A}$. As $\tilde{P} \to F$, the cone $\tilde{V} \to V_0$. Hence, the area of normal image of the cone $\tilde{V}$ and, therefore, the area of the normal image of the polyhedron $\tilde{P}$ becomes infinite. But this is impossible since the area of the normal image of the polyhedron is equal to $\int \varphi dx_1 \ldots dx_n < \infty$. This contradiction proves that the limit hypersurface F is projected in a unique manner onto the closed domain $\bar{G}$. Therefore, the boundary condition $z = \psi$ is satisfied on the boundary of the domain G. Thus the existence of a generalized solution of the Dirichlet problem is proved.

We shall now prove the uniqueness. Let z_1 and z_2 denote two different generalized solutions of Eq. (1) which coincide on the boundary of the domain G. Without loss of generality, we may assume that $z_1(x^0) < z_2(x^0)$ at some point x^0 of the domain G. Denote by G_0 the set of points of the domain G at which $z_1 < z_2$. Since the functions z_1 and z_2, being convex, are almost everywhere differentiable, there exists in G_0 a point x' at which z_1 and z_2 are both differentiable. Moreover, at this point $dz_1 \neq dz_2$. To see this, note that otherwise we would have $dz_1 = dz_2$ almost everywhere in G_0 and since $z_1 = z_2$ on the boundary of G_0, we would have $z_1 \equiv z_2$ in G_0, which is impossible. Displace the hypersurface F_1 a distance $z_2(x') = z_1(x')$ in the direction $z > 0$. Now the hypersurfaces F_1 and F_2 have a common point A' with a projection x' and different tangent hyperplanes at that point. Denote by F_1' the connected component of F_1, where $z_1 < z_2$. The point A' is a boundary point of F_1'. We denote by F_2' the corresponding domain on F_2 with the same projection onto the hyperplane $z = 0$. For each support hyperplane to F_2', there exists a parallel support hyperplane to F_1'. But for the support hyperplanes to F_1' at points close to A' there are no parallel support hyperplanes to F_2'. Since $\varphi > 0$, the area of the normal image of F_1' is greater than the area of the normal image of F_2'. But this is impossible since the domains F_1' and F_2' have a common projection onto the hyperplane $z = 0$ and hence the areas of the normal images must be the same. We have arrived at a contradiction. This completes the proof of the theorem.

3. *A priori* estimates for regular solutions.

Theorem 2. *Let $u(x)$ denote a convex regular solution of the equation*

$$\det (Z_{ij}) = \varphi(x) > 0, \; x \in G, \tag{1}$$

satisfying the boundary condition $u = 0$. Then at any interior point of G the second derivatives of the solution can be estimated in terms of the maximum of the absolute values of the solution and its first derivatives, the function $\varphi(x)$ and its derivatives up to second order, and the distance to the boundary of G.

PROOF. Consider the function

$$w = -ue^{u_\alpha^2/2} u_{\alpha\alpha}, \tag{2}$$

where the index α denotes differentiation in the fixed direction. It is obvious that the function w assumes its maximum at some point O of the domain G. By a unimodular transformation, we can make our choice of coordinate system in such a way that the direction α coincides with one of the axes and at the point O we have $u_{ij} = 0$ for $i \neq j$.

Take the logarithm of Eq. (2) and differentiate it twice with respect to x_i. Then, at the point O at which w attains its maximum, we have

$$\frac{u_{\alpha\alpha i}}{u_{\alpha\alpha}} + u_\alpha u_{\alpha i} + \frac{u_i}{u} = 0, \tag{3}$$

$$\frac{u_{\alpha\alpha ii}}{u_{\alpha\alpha}} - \left(\frac{u_{\alpha\alpha i}}{u_{\alpha\alpha}}\right)^2 + u_{\alpha i}^2 + u_\alpha u_{\alpha ii} + \frac{u_{ii}}{u} - \left(\frac{u_i}{u}\right)^2 = \frac{w_{ii}}{w}. \tag{4}$$

Multiplying equation (4) by $u_{\alpha\alpha}/u_{ii}$ and summing over i, we obtain, with the aid of Eq. (3),

$$\sum_i \frac{u_{ii\alpha\alpha}}{u_{ii}} - \sum_i \frac{(u_{\alpha\alpha i})^2}{u_{\alpha\alpha}u_{ii}} + n\frac{u_{\alpha\alpha}}{u} - \left(\frac{u_\alpha}{u}\right)^2 -$$
$$\sum_{i\neq\alpha} \frac{(u_{\alpha\alpha i})^2}{u_{\alpha\alpha}u_{ii}} + u_{\alpha\alpha}^2 + u_\alpha u_{\alpha\alpha} \sum_i \frac{u_{\alpha ii}}{u_{ii}} = \sum_i \frac{w_{ii}u_{\alpha\alpha}}{wu_{ii}}. \tag{5}$$

We now differentiate Eq. (1) twice with respect to x_α at the point O. We obtain successively

$$\sum_i \frac{u_{ii\alpha}}{u_{ii}} = \frac{\varphi_\alpha}{\varphi} = (\ln \varphi)_\alpha, \tag{6}$$

$$\sum_i \frac{u_{ii\alpha\alpha}}{u_{ii}} + \sum_{i\neq j}\left(\frac{u_{ii\alpha}u_{jj\alpha}}{u_{ii}u_{jj}} - \frac{u_{ij\alpha}^2}{u_{ii}u_{jj}}\right) = \frac{\varphi_{\alpha\alpha}}{\varphi}. \tag{7}$$

We now square Eq. (6) and subtract the result from Eq. (7). We obtain

$$\sum_i \frac{u_{ii\alpha\alpha}}{u_{ii}} - \sum_{i,j} \frac{(u_{ij\alpha})^2}{u_{ii}u_{jj}} = (\ln \varphi)_{\alpha\alpha}. \tag{8}$$

Subtracting Eq. (8) from Eq. (5) and observing that $w_{ii} \leqslant 0$ at the point O and that

$$-\sum_i \frac{(u_{\alpha\alpha i})^2}{u_{\alpha\alpha}u_{ii}} - \sum_{i\neq\alpha} \frac{(u_{\alpha\alpha i})^2}{u_{\alpha\alpha}u_{ii}} + \sum_{i,j} \frac{(u_{ij\alpha})^2}{u_{ii}u_{jj}} \geqslant 0,$$

we obtain

$$\frac{nu_{\alpha\alpha}}{u} - \left(\frac{u_\alpha}{u}\right)^2 + u_{\alpha\alpha}^2 + u_\alpha u_{\alpha\alpha}(\ln \varphi)_\alpha + (\ln \varphi)_{\alpha\alpha} \leqslant 0. \tag{9}$$

Multiplying inequality (9) by $u^2e^{u_\alpha^2}$ and introducing w, we obtain

$$w^2 + Aw + B \leqslant 0, \tag{10}$$

where A and B admit estimates depending on the quantities indicated in the theorem. From inequality (10) one can easily obtain an estimate for w. If we denote it by w_0, then, everywhere in the domain G,

$$u_{\alpha\alpha} \leqslant \frac{w_0}{|u|}. \tag{11}$$

Since the function $u(x)$ is convex and vanishes on the boundary of the domain G,

$$\frac{|u|}{\delta} \geqslant \frac{\max|u|}{D},$$

where δ is the distance form the point in question to the boundary of G and D is the diameter of the domain G. Hence, it follows from (11) that

$$u_{\alpha\alpha} \leqslant \frac{w_0 D}{\delta \max |u|}. \tag{12}$$

Since α was chosen arbitrarily, the estimate (12) holds for any direction. This completes the proof of the theorem.

Theorem 3. *Let $u(x)$ denote a convex solution of Eq.* (1) *in the domain G that vanishes on the boundary of that domain. Then at any interior point of G, this solution and its derivatives up to third order can be estimated in terms of the right-hand side φ of the equation, the derivatives of φ up to third order, and the distance from the point to the boundary of G.*

PROOF. Let F denote the convex hypersurface defined by the solution $u(x)$. Denote by x_0 the point of the domain G at which the absolute value $|u|$ assumes its maximum $|u_0|$. We denote by A_0 the corresponding point of the hypersurface F. Let ω_0 denote a ball having center at x_0 on the hyperplane $z = 0$ and containing the domain G. Let V_0 denote a cone with vertex A_0 and base ω_0. It is obvious that the normal image of the cone is contained in the normal image of the hypersurface F. Hence,

$$\varkappa \left(\frac{|u_0|}{R}\right)^n \leqslant \int_G \varphi \, dx_1 \ldots dx_n,$$

where R is the radius of the ball ω_0 and $\varkappa$ is the volume of a unit ball. From this we obtain an estimate for

$$|u_0| = \max |u(x)|.$$

Let us estimate the first derivatives of the solution at an arbitrary point x' of the domain G. Let A' be a point of the hypersurface F whose projection is x'. In the hyperplane $z = 0$, consider the ball ω' of radius δ with center at x', where δ is the distance from the point x' to the boundary of the domain. Obviously, the slope of the tangent hyperplane to F at the point A' is no greater than the slope of the tangent hyperplanes to the cone V' with vertex A' and base ω'. Thus,

$$|u_\alpha| \leqslant \frac{|u(x')|}{\delta} < \frac{\max|u|}{\delta}.$$

Since an estimate for max $|u|$ has already been obtained, this inequality gives an estimate for $|u_\alpha|$.

To estimate the second derivatives, we consider the solution $u(x)$ in a subdomain G_ε of the domain G such that

$$u < -\varepsilon, \qquad \varepsilon > 0.$$

In G_ε, we have uniform estimates for the solution and its first derivatives. By Theorem 2, it follows from this that there exists an estimate for the second derivatives $u_{\alpha\beta}$ inside the domain G_ε. Since ε is arbitrary, we obtain an estimate everywhere inside the domain G.

Finally, to estimate the third derivatives, we again consider the solution in the domain G_ε. Inside this domain, we have uniform estimates for the solution and its derivatives up to second order. By Theorem 3 of Sec. 3, we can find an estimate for the third derivatives inside the domain G_ε. And since ε is arbitrary, we obtain an estimate everywhere inside G. The theorem is proved.

4. Regularity of the generalized solutions.

Theorem 4. *The generalized solution $u(x)$ of the equation*

$$\det(z_{ij}) = \varphi(x) > 0, \; x \in G, \tag{1}$$

satisfying the linear boundary condition

$$u(x) = a_1x_1 + \ldots + a_nx_n + a_0, \tag{2}$$

is regular if the right-hand side $\varphi(x)$ of the equation is regular. If φ is of class C^k, where $k \geqslant 3$, then the solution is of at least class $C^{k+1,\alpha}$, where $0 < \alpha < 1$. If φ is analytic, the solution is analytic.

PROOF. Without loss of generality, one may assume that the first derivatives of the solution $u_\alpha(x)$ are uniformly bounded in the domain G. This is always the case if we consider the solution in the domain G_ε defined by

$$u(x) - (a_1x_1 + \ldots + a_nx_n + a_0) < -\varepsilon, \qquad \varepsilon > 0.$$

We also suppose that in the domain G

$$u - \sum x_\alpha p_\alpha \geqslant 0,$$

where the p_α are the slopes of the support hyperplanes to the hypersurface defined by the solution. To attain this, we need only consider the solution $u - c$ for sufficiently large c.

Let us displace the domain G in the direction $z > 0$ into the hyperplane $z = 1$. In this position, we denote it by G'. We denote by Ω the unit hypersphere $z^2 + x_1^2 + \ldots + x_n^2 = 1$. We denote the cone projecting the domain G' from the coordinate origin by V' and we denote its intersection with the hypersphere by Ω'. Inside the cone V', we define the convex function positively homogeneous of first degree,

$$H'(z, x_1, \ldots, x_n) = zu\left(\frac{x_1}{z}, \ldots, \frac{x_n}{z}\right).$$

We extend the function H' from the cone V' to the whole space, setting

$$H'(\bar{z}, \bar{x}) = \max_{x \subset G} \left| \bar{x}_\alpha p_\alpha + \bar{z}\left(u - \sum x_\alpha p_\alpha\right) \right|.$$

This function is symmetric with respect to the coordinate origin and convex. Inside the cone V', it coincides with the function $H'(z, x)$. We denote $\overline{F}$ a closed convex hypersurface having H as its support function. We denote by F' the part of the hypersurface which has the domain Ω' as its spherical image.

Let us show that the Gaussian curvature of the hypersurface F' at the point with exterior normal of direction $x_1, x_2, \ldots, 1$ is

$$K = 1/(1 + x_1^2 + \ldots + x_n^2)^{\frac{n}{2}+1} \varphi.$$

To do this, we approximate the function $u(x)$ with a regular convex function $\widetilde{u}(x)$. We now construct a support function $\widetilde{H}(z, x)$, a hypersurface $\widetilde{F}$, and a domain $\widetilde{F}'$ on $\widetilde{F}$ in a manner similar to that in which H, F, and F' were constructed. Let us write det $(\widetilde{u}_{ij}) = \widetilde{\varphi}(x)$. Consider the following set functions

$$s = \int_M \varphi \, dx_1 \ldots dx_n, \quad \bar{s} = \int_M \widetilde{\varphi} \, dx_1 \ldots dx_n,$$

$$\sigma = \int \varphi (1 + x_1^2 + \ldots + x_n^2)^{1/2} dx_1 \ldots dx_n,$$

$$\tilde{\sigma} = \int \widetilde{\varphi} (1 + x_1^2 + \ldots + x_n^2)^{1/2} dx_1 \ldots dx_n.$$

When $\widetilde{u} \to u$, then $\widetilde{F} \to F$ and $\widetilde{s}$ converges weakly to s. From this it follows that $\widetilde{\sigma}$ converges weakly to σ. But $\widetilde{\sigma}$ is the surface function of the hypersurface $\widetilde{F}'$. Since the hypersurface $\widetilde{F}'$ converges to F', the limit of $\widetilde{\sigma}$, i.e., the function σ, is the surface function of the hypersurface F'. The Gaussian curvature is the limit of the ratio of the area of the spherical image to the area of the hypersurface. Therefore, the Gaussian curvature of the hypersurface F' is

$$K = 1/(1 + x_1^2 + \ldots + x_n^2)^{\frac{n}{2}+1} \varphi.$$

Construct now a sequence of positive functions $K_m(\xi)$ of class C^k on the unit hypersurface Ω with the following properties:

1. The function $K_m(\xi)$ is symmetric, i.e., $K_m(\xi) = K_m(-\xi)$.
2. In the domain Ω', $K_m(\xi)$ coincides with the Gaussian curvature of the hypersurface F' at a point with exterior normal ξ.
3. The set function defined by the formula

$$\sigma_m = \int \frac{d\omega}{K_m(\xi)},$$

converges weakly to a surface function of the hypersurface F.

Construct a closed convex hypersurface F_m with Gaussian curvature K_m (*Minkowski's problem*). This is possible since the condition

$$\int_{\Omega} \xi \frac{d\omega}{K_m(\xi)} = 0$$

holds by virtue of the symmetry of the function $K_m(\xi)$. The hypersurface F_m is of at least class C^4. As $m \to \infty$, the hypersurfaces F_m converge to the hypersurface F because of the uniqueness of the solution of the Minkowski problem. Let H_m denote the support function of the hypersurface F. Set

$$u_m(x_1, \ldots, x_n) = H_m(1, x_1, \ldots, x_n).$$

Since the Gaussian curvature of the hypersurface F_m in the domain with spherical image Ω' is equal to $1/(1 + x_1^2 + \ldots + x_n^2)^{\frac{n}{2}+1}$, the function $u_m(x)$ satisfies Eq. (1) and converges to the generalized solution $u(x)$ as $m \to \infty$. By Theorem 3, *a priori* estimates can be given for $u_m(x) - (a_1x_1 + \ldots + a_nx_n + a_0)$. Therefore, the limit function $u(x)$ is of class C^2. Further regularity of $u(x)$ is a consequence of the general theorem on regularity of solutions of elliptic equations. The theorem is proved.

Theorem 5. *Any strictly convex solution of Eq.* (1) *with regular right-hand side* φ *is regular. If the function* $\varphi(x)$ *is of class* C^k, *where* $k \geqslant 3$, *then the solution is of class* $C^{k+1,\alpha}$, *where* $0 < \alpha < 1$. *If the function* φ *is analytic, so is the solution.*

PROOF. Let F denote the hypersurface defined by the solution, A an arbitrary point of F, and α a support hyperplane at the point A. We displace α a small distance in the direction $z > 0$ so that the edge of the hypersurface F remains over the hyperplane. Suppose that the equation of the displaced hyperplane is $z = a_1x_1 + \cdots + a_nx_n + a_0$. Then the solution $u(x)$ in the domain G defined by the inequality

$$u(x) - \sum a_\alpha x_\alpha - a \leqslant 0,$$

satisfies the conditions of Theorem 4 and hence is regular at the point A. The theorem is proved.

Theorem 6. *A complete convex hypersurface F satisfying Eq.* (1) *is regular without the assumption of strict convexity.*

PROOF. First of all, we note that the hypersurface F has points of strict convexity. Indeed, the set of directions of support hyperplanes having more than one common point with the hypersurface is of measure zero. Therefore, if the hypersurface F has no points of strict convexity, its normal image must have zero area. But for the hypersurface F, this area is positive since $\varphi > 0$.

Thus, the hypersurface F has points of strict convexity. We take one such point A and at this point we consider a support hyperplane α that has no other points in common with F. Let B denote an arbitrary point of the hypersurface F. Displace the hyperplane α in the direction $z > 0$ so that in this position the point B is under the hyperplane α. Now we apply Theorem 4 to the solution $u(x)$ in the domain G defined by the inequality

$$u - \left(\sum a_k x_k + a_0\right) \leqslant 0.$$

By this theorem, the solution $u(x)$ is regular in the domain G. Hence, the hypersurface F is regular at an arbitrary point B. The theorem is proved.

5. An example of a generalized solution which is not regular. Let us compute the Hessian of the function

$$z(x_1, \ldots, x_n) = u\left(x_1, \left(\sum_{k>1} x_k^2\right)^{1/2}\right).$$

Let us set

$$x_1 = \xi, \qquad \Big(\sum_{k>1} x_k^2\Big)^{1/2} = \eta.$$

By straightforward computations one can verify that

$$\det\,(z_{ij}) = (u_{\xi\xi}u_{\eta\eta} - u_{\xi\eta}^2)\,(u_\eta/\eta)^{n-2}. \tag{1}$$

Let u denote the function

$$u = (1 + \xi^2)\eta^\alpha.$$

Then

$$\det\,(z_{ij}) = (2\alpha)^{n-1}\,(1 + \xi^2)^{n-2}\,\eta^{n(\alpha-2)+2}\,(\alpha - 1 - (\alpha + 1)\,\xi^2).$$

Now set

$$\alpha = 2 - \frac{2}{n}.$$

Then

$$n\,(\alpha - 2) + 2 = 0$$

and

$$\det\,(z_{ij}) = (2\alpha)^{n-1}\,(1 + x_1^2)^{n-2}\,(\alpha - 1 - (\alpha + 1)\,x_1^2).$$

Set

$$\varphi\,(x_1, \ldots, x_n) = (2\alpha)^{n-1}\,(1 + x_1^2)^{n-2}\,(\alpha - 1 - (\alpha + 1)\,x_1^2).$$

Let us show that the function

$$z\,(x_1, \ldots, x_n) = (1 + x_1^2)\Big(\sum_{k>1} x_k^2\Big)^{1-\frac{1}{n}}$$

is a generalized solution of the equation

$$\det\,(z_{ij}) = \varphi \tag{2}$$

in the ball Ω: $x_1^2 + \ldots + x_n^2 \leqslant \varepsilon^2$ of sufficiently small radius ε. First of all, we observe that the function $z(x)$ is analytic everywhere except on the x_1 axis. At each point which does not belong to the x_1 axis, $d^2 z > 0$. Indeed the kth-order minor (where $k \geqslant 2$) of the matrix (z_{ij}) is

$$\Delta_k = (2\alpha)^{k-1}(1 + x_1^2)^{k-2}(\alpha - 1 - (\alpha + 1)x_1^2) > 0,$$

$$\Delta_1 = 2\left(\sum_{k>1} x_k^2\right)^{1-\frac{1}{n}} > 0.$$

By Jacobi's theorem, the quadratic form $d^2 z$ is positive-definite. Hence, the function $z(x)$ is convex in a neighborhood of every point not on the x_1 axis. Since the function $z(x)$ is smooth,* it is convex in the large.

Since the function $z(x)$ is smooth everywhere and regular** everywhere except on the x_1 axis, the area of the normal image of the hypersurface defined by $z(x)$ is given by

$$s = \int \varphi \, dx_1 \ldots dx_n,$$

and, hence, the function $z(x)$ is a generalized solution of Eq. (2) with positive analytic right-hand side φ. But as was mentioned above, $z(x)$ is not twice differentiable on the x_1 axis. This example justifies the condition of strict convexity of the solution in Theorem 5. In this example, that condition is violated on the x_1 axis.

This same example of a generalized solution of Eq. (2) can be treated from a different point of view. Let ψ denote the restriction of $z(x)$ to the boundary of the ball Ω: $x_1^2 + x_2^2 + \ldots + x_n^2 \leqslant \varepsilon^2$. By Theorem 1, the Dirichlet problem for Eq. (2) has a generalized solution which coincides with ψ on the boundary of Ω. Since the solution is unique, it coincides with $z(x)$. This

*Editor's note. Of class C^1.

**Editor's note. Of class C^k, where $k \geqslant 2$.

justifies the requirement that the boundary values of a regular solution of the Dirichlet problem be given by a function of class C^2 (see subsection 6). In the above example, ψ is of class $C^{1,1-\frac{2}{n}}$.

In connection with the regular solution of the Dirichlet problem which will be given in subsection 6, let us construct one more example of the solution of Eq. (2) with a positive constant in the right-hand side.

Lemma 2. *Let b, k_0, η_0, and A denote positive numbers and let $\xi(x)$ denote a solution of the equation*

$$\xi'' + k_0\xi^{n-1} = 0 \qquad (n > 2)$$

on the interval $-b \leqslant x \leqslant b$. Suppose that $\xi(x)$ is convex in the direction $\xi > 0$ and equal to zero at the end-points of the interval. Then, the equation

$$\det(z_{ij}) = k_0 \tag{3}$$

has, on the domain G_1 defined by

$$\frac{1}{\xi(x_1)}\Big(\sum_{k>1} x_k^2\Big)^{1/2} \leqslant \eta_1 \leqslant \eta_0,$$

a convex solution $z_1(x_1, \ldots, x_n)$ that satisfies the inequality

$$z_1(x) \geqslant A \sum_{k>1} x_k^2,$$

on the boundary of G_1 and is negative inside that domain.

PROOF. We look for the solution $z_1(x)$ in the form

$$z = \xi(x_1)\, u\left(\frac{1}{\xi(x_1)}\Big(\sum_{k>1} x_k^2\Big)^{1/2}\right).$$

Set

$$\eta = \frac{1}{\xi(x_1)}\Big(\sum_{k>1} x_k^2\Big)^{1/2}.$$

Then

$$\det(z_{ij}) = -\frac{\xi''}{\xi^{n-1}} u''(u'\eta - u)\left(\frac{u'}{\eta}\right)^{n-2} = k_0.$$

Since $\xi'' + k_0\xi^{n-1} = 0$, we obtain for $u(\eta)$ the equation

$$u''(u'\eta - u)(u'/\eta)^{n-2} = 1. \tag{4}$$

Let us consider a solution of this equation for $\eta \geqslant 0$ such that $u(0) = \delta < 0$, $u'(0) = 0$, and $u''(0) > 0$. Since u'' has the same sign everywhere, $u(\eta)$ is a convex function.

Let a and η_0 denote positive constants. We assert that there exists $\eta' \leqslant \eta_0$ such that, for sufficiently small $|\delta|$, the function $a\eta^2 - u(\eta)$ vanishes when $\eta = \eta'$. Suppose the assertion is not true. This means that $u \leqslant a\eta^2$ when $\eta \leqslant \eta_0$ no matter how small $|\delta|$ is. Passing to the limit as $\delta \to 0$, we obtain a function $u_0(\eta)$ with the properties that $u_0(0) = 0$ and $u_0(\eta) \leqslant a\eta^2$ for $\eta < \eta_0$. The function u_0 satisfies Eq. (4). Since the convex curve $u = u_0(\eta)$ lies between the η axis and the parabola $u = a\eta^2$, we have $u_0'\eta - u_0 \leqslant A\eta^2$, where A is a constant depending only on a. Remembering that $u_0 \leqslant a\eta^2$, we conclude that u_0'/η is bounded. Since $u_0 \leqslant a\eta^2$, by choosing η sufficiently small, we can have $u_0'' \leqslant 2a$. Also, for sufficiently small values of η, the expression

$$u_0''(u_0'\eta - u_0)(u_0'/\eta)^{n-2}$$

can be made arbitrarily small, a fact that contradicts Eq. (4). Thus, $a\eta_1^2 - u(\eta_1) = 0$ for some $\eta_1 \leqslant \eta_0$.

Denote by G_1 the domain defined by the inequality

$$\frac{1}{\xi(x_1)}\left(\sum_{k>1} x_k^2\right)^{1/2} \leqslant \eta_1.$$

On the boundary of the domain G_1, the function $z_1(x)$ satisfies the inequality

$$z_1(x) = \xi a\eta_1^2 = \frac{a}{\xi}\left(\sum_{k>1} x_k^2\right) \geqslant \frac{a}{\xi(0)} \sum_{k>1} x_k^2.$$

Since a is arbitrary, one can set $a/\xi(0) = A$. Thus, the so-constructed function $z_1(x)$ satisfies Eq. (3) in the domain G_1, and, on the boundary of G_1,

$$z_1(x) \geqslant A \sum_{k>1} x_k^2.$$

For $x_2 = x_3 = \ldots = x_n = 0$ inside the domain G_1, the function z_1 is negative. The convexity of $z_1(x)$ can be proved in a manner similar to that of the previous example. This completes the proof of the lemma.

6. A regular solution of the Dirichlet problem. It was shown in subsection 1 (Theorem 1) that, for a strictly convex domain G, a continuous positive function $\varphi(x)$ defined on G, and an arbitrary continuous function ψ defined on the boundary of G, the Dirichlet problem for the equation

$$\det(z_{ij}) = \varphi(x) > 0$$

has a generalized solution. If the function φ is sufficiently regular and the generalized solution is strictly convex, then the latter is regular (Theorem 5). Unfortunately, the condition of strict convexity applies the solution and not to the data of the problem. In connection with this, the following theorem is of interest:

Theorem 7. *The Dirichlet problem for Eq.* (1) *with a regular right-hand side* $\varphi(x)$ *always has, in a strictly convex domain G, a regular solution if the boundary of G and the boundary values of the solution are of class* C^2. *The solution is of class* $C^{k+1,\alpha}$, *where* $0 < \alpha < 1$, *if the function* φ *is of class* C^k *(where* $k \geqslant 3$*). The solution is analytic if the function* φ *is analytic.*

PROOF. Let $z(x)$ denote a generalized solution of Eq. (1) in the domain G with the boundary values ψ. Denote by F the convex hypersurface defined by the solution $z(x)$. If the hypersurface F is strictly convex, then the solution $z(x)$ is regular (by Theorem 5). Suppose the hypersurface is not strictly convex. Then there exists a support hyperplane σ of F such that its intersection with F is a convex set M containing more than one point. We assert that each point of strict convexity of the set M belongs to the boundary F.

Let A be a point of strict convexity not on the boundary of F. Consider a support hyperplane α to M at the point A and suppose that A is the only common point of α and M. The hyperplane α divides σ into two parts. Let σ_1 denote the part which does not contain M. Let F_1 denote the part of the hypersurface F which projects into σ_1. The hypersurface F_1 is situated above σ_1. Displace the hyperplane α a small distance toward M into the position α'. Rotate the hyperplane σ about α' through a small angle into the position σ' so that the point A is below the hyperplane σ'. Then F will be partitioned by σ' into two components. Let F_A be the component of F containing the point A. When α' is sufficiently close to α and σ' is sufficiently close to σ, the boundary of F_A is separated from the boundary of F. By Theorem 4, the hypersurface F_A is regular in a neighborhood of the point A. But then A is a point of strict convexity of F and it cannot belong to M. We have arrived at a contradiction. Thus, the points of strict convexity of the set M lie on the boundary of F. Hence, there exists on F a linear segment g whose end-points belong to the boundary of F.

Let $\sigma\colon z = a_1x_1 + \ldots + a_nx_n + a_0$ denote a support hyperplane of F along the segment g. The function $\bar{z}(x) = z(x) - (a_1x_1 + \ldots + a_nx_n + a_0)$ also satisfies Eq. (1). The boundary values of $\bar{z}(x)$ are also of class C^2. Because of this, we can, without loss of generality, assume that the segment g lies on the x_1 axis, that its midpoint coincides with the coordinate origin, and that the coordinate hyperplane $z = 0$ is the support function for the hypersurface F defined by the solution.

Let B_1 and B_2 denote the end-points of the segment g. In a neighborhood of the point B_1, the boundary of the domain G is defined by the equation $x_1 = \chi_1(x_2, \ldots, x_n)$. The analogous equation $x_1 = \chi_2(x_2, \ldots, x_n)$ defines the boundary of G in a neighborhood of B_2. The boundary of the hypersurface F in the neighborhood of the point B_1 is defined by the two equations

$$x_1 = \chi_1(x_2, \ldots, x_n), \qquad z = \psi_1(x_2, \ldots, x_n).$$

Similarly, the boundary of F is defined in the neighborhood of B_2 by the equations

$$x_1 = \chi_2 (x_2, \ldots, x_n), \qquad z = \psi_2 (x_2, \ldots, x_n).$$

Since the coordinate hyperplane $z = 0$ is a support hyperplane to F at the points B_1 and B_2, we have $\psi_1 = 0$, $d\psi_1 = 0$ and, correspondingly, $\psi_2 = 0$, $d\psi_2 = 0$. By the assumption of the theorem, the function ψ defining the boundary values of the solution is of class C^2. Therefore, in the neighborhoods of B_1 and B_2, i.e., for sufficiently small values of $\sum_{k>1} (x_k)^2$, the following inequalities hold:

$$\psi_1 < A \sum_{k>1} x_k^2, \qquad \psi_2 < A \sum_{k>1} x_k^2,$$

where A is a positive number. Denote by Φ_1 the hypersurface defined by the equation

$$z = A \sum_{k>1} x_k^2.$$

Construct the convex hull of the boundary of F and denote by Φ the part of it that is convex in the direction $z < 0$. Let $\zeta(x)$ denote the convex function defining Φ. It is easy to see that

$$z(x) \leqslant \zeta(x),$$

where $z(x)$ is the function defining F. The hypersurface Φ_1 in a neighborhood of the segment g is situated above the convex hull Φ. Hence, in some neighborhood of the segment g we have

$$z(x) \leqslant \zeta(x) \leqslant A \sum_{k>1} x_k^2.$$

Consider now the solution $z_1(x)$ of the equation

$$\det(z_{ij}) = k_0 < \varphi(x),$$

which appeared in Lemma 2 (subsection 5). On the boundary of the domain G_1 defined by

$$\frac{1}{\xi(x_1)}\left(\sum_{k>1} x_k^2\right)^{1/2} \leqslant \eta_1,$$

we have

$$z_1(x) \geqslant A \sum_{k>1} x_{2k}.$$

Hence, on the boundary of the domain G_1,

$$z(x) \leqslant z_1(x).$$

Consider now the convex hypersurfaces F' and F_1' determined by the functions $z(x)$ and $z_1(x)$, respectively, in the domain G_1. The functions $z(x)$ and $z_1(x)$ have the following property: At the coordinate origin, we have $z_1 < z$ and, on the boundary of the domain G_1, we have $z_1 \geqslant z$. Denote by G_1' the connected component of the domain G_1 (where $z_1 < z$) which contains the coordinate origin. Obviously, $z_1 = z$ on the boundary of G_1'. Denote by F'' and F_1'' the parts of the hypersurfaces F' and F_1' which project into the domain G_1'. Since $z = z_1$ on the boundary of the domain G_1' and $z_1 < z$ inside the domain, the normal image of the hypersurface F'' is contained in the normal image of F_1''. Hence,

$$\int_{G_1'} \varphi(x)\, dx_1 \ldots dx_n \leqslant \int_{G_1'} k_0\, dx_1 \ldots dx_n.$$

But this is impossible since $\varphi > k_0$. We have arrived at a contradiction. The theorem is proved.

§6. ON IMPROPER CONVEX AFFINE HYPERSPHERES

An *improper affine hypersurface* is a complete hypersurface (in $(n + 1)$-dimensional Euclidean space) defined in a suitably chosen

Cartesian coordinate system $x_1, \ldots, x_n, z$ by an equation of the form $z = z(x_1, \ldots, x_n)$, where $z(x)$ is a function satisfying the differential equation

$$\det (z_{ij}) = \text{const} > 0.$$

The subject of this section is the proof of the following theorem:

Theorem. *A convex improper affine hypersurface is an elliptic paraboloid.*

This result was obtained by K. Jörgens [9] for $n = 2$ and by E. Calabi [10] for $n = 3, 4, 5$. We shall prove it here for arbitrary n.

1. On the location of the center of gravity of a convex body. In this subsection, we shall treat some properties of the center of gravity of a convex body in n-dimensional Euclidean space. These properties are well-known and we discuss them here to give a complete exposition of the problem.

Let S denote the center of gravity of a convex body and let h_1 and h_2 denote the respective distances from S to two parallel support hyperplanes. Then

$$\frac{1}{n} \leqslant \frac{h_1}{h_2} \leqslant n.$$

To see this, let α_1 and α_2 denote the parallel support hyperplanes and let α denote a hyperplane parallel to them and passing through the center of gravity S (Fig. 1). Construct a cone with vertex A_1, which projects the section of the body by the hyperplane α. If one replaces the body with the cone, the center

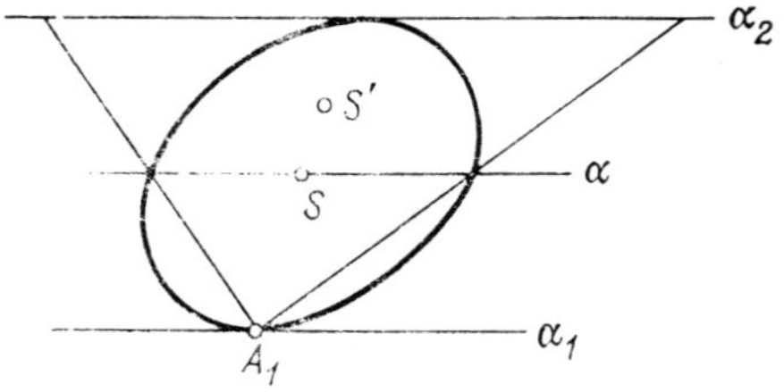

FIG. 1

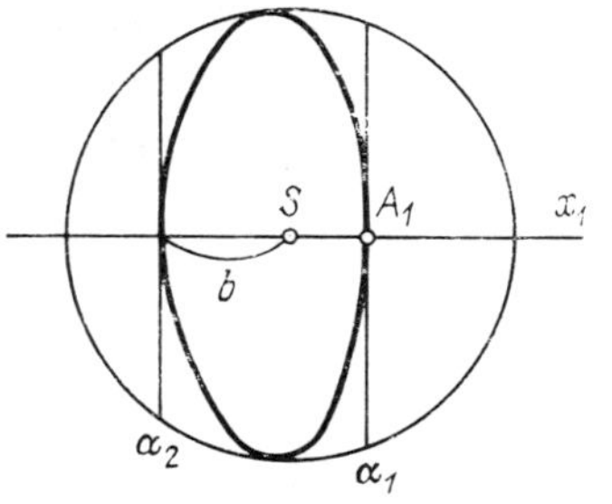

FIG. 2

of gravity is displaced toward the hyperplane α_2. For the center of gravity S' of the cone, we have $h_1'/h_2' = n$. Therefore, for the given body, the relation is $h_1/h_2 \leqslant n$. In a similar way, one can prove that $h_2/h_1 \leqslant n$, i.e., $h_1/h_2 \geqslant 1/n$. The assertion is proved.

Let T denote a convex body with center of gravity S and let E denote the ellipsoid of smallest volume with center at S that contains the body. Then the body T contains the ellipsoid homothetic to E with respect to the center S with coefficient of homothety $1/n^{3/2}$. Without loss of generality, one can assume the ellipsoid E to be a hypersphere. (The general case can be reduced to this particular one by a suitable affine transformation.) Let r denote the radius of the hypersphere E. Suppose the assertion is not true. Then consider that point A_1 of the boundary of the body T that is closest to S; obviously, the distance between S and A_1 is less than $r/n^{3/2}$. Let α_1 denote support hyperplane to T passing through the point A_1, and let α_2 denote the parallel support hyperplane (Fig. 2). The distance between α_2 and S is equal to $b < r/\sqrt{n}$.

Introduce the Cartesian coordinates x_i, taking the point S for the coordinate origin and directing x_1 axis normally to the hyperplanes α_1 and α_2. Since the hypersphere E contains the body T and since T lies between the hyperplanes α_1 and α_2, we have, for each point x of T,

$$|x_1| < b, \qquad x_1^2 + x_2^2 + \ldots + x_n^2 \leqslant r^2.$$

Consider the ellipsoid E':

$$\frac{x_1^2}{\alpha^2} + \frac{1}{\beta^2} \sum_{k>1} x_k^2 = 1,$$

where

$$\alpha = b\sqrt{n}, \qquad \beta = \left(\frac{n(r^2 - b^2)}{n-1} \right)^{1/2}.$$

We assert that the ellipsoid E' also contains the body T and that the volume of E' is less than the volume of the hypersphere E. Indeed, since $b < r/\sqrt{n}$, we have $\alpha < \beta$. Substituting the coordinates of an arbitrary point x of the body into the equation of the ellipsoid E' we obtain

$$\frac{x_1^2}{\alpha^2} + \frac{1}{\beta^2} \sum_{k>1} x_k^2 < b^2 \left(\frac{1}{\alpha^2} - \frac{1}{\beta^2} \right) + \frac{r^2}{\beta^2} = 1.$$

This means that the body T belongs to E'. The ratio of the volume of E' to the volume of E is

$$\frac{V'}{V} = \frac{\alpha\beta^{n-1}}{r^n} = \sqrt{n}\,\frac{b}{r} \left(\frac{n\left(1 - \frac{b^2}{r^2}\right)}{n-1} \right)^{\frac{n-1}{2}}$$

Noting that $b/r < 1/\sqrt{n}$, we get $V'/V < 1$. But this contradicts the assumption that the hypersphere E has the least volume among all the ellipsoids with center at S that contain the body. The assertion is proved.

2. Special sections of affine hyperspheres. Let an affine hypersphere F be defined by the equation $z = z(x)$ in Cartesian coordinates $x_1, \ldots x_n, z$. Without loss of generality, one can assume that the origin O of the coordinate system lies on F. Let T denote the convex body bounded by F. Consider an arbitrary point S on the positive semiaxis z (inside the body T) and a hyperplane σ passing through S. Let T_σ denote the section of T by the hyperplane σ and let $\bar{T}_\sigma$ denote its projection onto the hyperplane $z = 0$. We call the section of the body T by σ *central* if the point S is the center of gravity of T_σ; correspondingly, the origin O of the coordinate system is the center of gravity of the projection $\overline{T}_\sigma$. Suppose that

$$z = p_1 x_1 + \dots + p_n x_n + h$$

is the equation of the hyperplane σ passing through the point S. The coordinates of the center of gravity of the body $\overline{T}_\sigma$ are functions of the coefficients in the equation of the hyperplane σ:

$$\begin{aligned} z &= f(p_1, p_2, \dots, p_n, h), \\ x_1 &= f_1(p_1, p_2, \dots, p_n, h), \\ &\dots\dots\dots\dots\dots \\ x_n &= f_n(p_1, p_2, \dots, p_n, h). \end{aligned}$$

The assertion that a central section exists is equivalent to the assertion that the system

$$\left.\begin{aligned} f_1(p_1, p_2, \dots, p_n, h) &= 0, \\ f_2(p_1, p_2, \dots, p_n, h) &= 0, \\ &\dots\dots\dots \\ f_n(p_1, p_2, \dots, p_n, h) &= 0 \end{aligned}\right\} \qquad (*)$$

has a solution for every h.

Let E denote the set of points S on the semi-axis $z \geqslant 0$ through which pass central sections. It is obvious that the point O belongs to E. Therefore, E is not empty. We shall show that the set E is both open and closed. Since E is not empty, this will mean that all points of the semi-axis $z \geqslant 0$ belong to E.

Suppose the system (*) is solvable when $h = h_0$. Let us show that it is solvable for all h sufficiently close to h_0. Consider a central section σ_0 passing through the point $S_0(h_0)$ and an arbitrary hyperplane σ which passes through the point S_0 and is close to σ_0. The intersection of the hyperplanes σ and σ_0 is an $(n-2)$-dimensional plane α. Let $\bar{\alpha}$ denote its projection onto the hyperplane $z = 0$. This $\bar{\alpha}$ partitions the body $\overline{T}_{\sigma_0}$ (see Fig. 3) into parts A_1^0 and A_2^0 and the body $\overline{T}_\sigma$ into parts A_1 and A_2. Since the hypersurface F is convex, either $A_1^0 \subset A_1$ and $A_2^0 \supset A_2$ or $A_1^0 \supset A_1$ and $A_2^0 \subset A_2$. In either case, the shift from σ_0 to σ results in a displacement of the center of gravity O. Moreover, one can easily show that if dp is the change in the angular coefficients in the shift from σ_0 to σ and dx is the displacement of the center of gravity, then $|dx| \geqslant c\ |dp|$, where c is a positive constant. Since $|dx| \geqslant c\ |dp|$, the Jacobian of the system (*) is nonzero for the values of p and h corresponding to the hyperplane σ_0. Hence, the system (*) is solvable for values of h close to h_0. Thus, if the point S_0 belongs to the set E, then E also contains points close to S_0; i.e., E is open.

We shall prove now that the set E is closed.

Let S_0 denote a boundary point of E. Let S_k be a sequence of points of E converging to the point S_0. Suppose that σ_k is a central section passing through the point S_k. Without loss of generality, we can assume that the sequence of hyperplanes σ_k converges to a certain hyperplane σ_0. We need to prove that the section by the hyperplane σ_0 is a central one. This is obvious when the body T_{σ_0} is bounded. Suppose that T_{σ_0} is unbounded. According to the first property of the center of gravity (subsection 1) applied to the bodies T_{σ_k}, the body contains a straight line. Then its surface must be a cylinder. This is impossible since

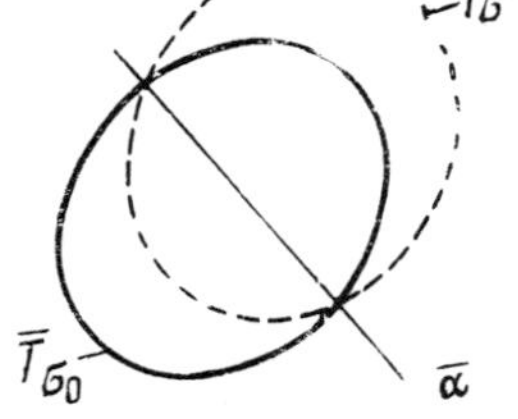

FIG. 3

the section of the strictly convex hypersurface F by the hyperplane σ_0 has to be strictly convex. Therefore, the body T_{σ_0} is bounded and hence the section by the hyperplane σ_0 is a central section. Thus, the boundary of the set E is contained in E; i.e., the set E is closed. This completes the proof of the theorem.

3. Some size relations for an affine hypersphere. Suppose that a convex affine hypersphere satisfies the equation

$$\det (z_{ij}) = 1. \tag{1}$$

Without loss of generality, one can assume that the coordinate origin O belongs to the hypersphere F. We preserve the notation of subsection 2 and, at the point $S(h)$ on the positive semi-axis z, we consider the central section of the body T bounded by the hypersphere F. Let T_σ denote the intersection of the hyperplane σ and the body T. Let $\overline{T}_\sigma$ denote the projection of T_σ onto the hyperplane $z = 0$. Construct the ellipsoid E of least volume with center at O that contains the body $\overline{T}_\sigma$. We transform the ellipsoid E into a hypersphere by an equiaffine transformation of the hyperplane $z = 0$. Let r denote the radius of this hypersphere. We assert that there exist positive constants c_1 and c_2, depending only on n such that

$$c_1 \leqslant \frac{r^2}{h} \leqslant c_2.$$

Extend the equiaffine transformation of the hyperplane $z = 0$ to the whole space, taking $z' = z$. Then the hypersurface F is transformed into a hypersurface F' which also satisfies Eq. (1). Suppose that

$$z = p_1x_1 + \dots + p_nx_n$$

is the equation of the tangent hyperplane of F' at the point O. The affine transformation

$$x_1' = x_1,\ x_2' = x_2,\ \dots,\ x_n' = x_n,$$
$$z' = p_1x_1 + p_2x_2 + \dots + p_nx_n$$

maps the hypersurface F' into the hypersurface F'', which also satisfies Eq. (1). The section of F'' by the hyperplane $z = h$ is a central one; the minimal ellipsoid containing its projection is a hypersphere. To avoid introducing new notation, let us assume that the hypersurface F already possesses the properties indicated. Specifically, the hyperplane σ of the central section passing through the point $S(h)$ is defined by the equation $z = h$, and the ellipsoid of minimal volume containing the body $\overline{T}_\sigma$ is a hypersphere.

Let M be an arbitrary Borel set of points on the hypersurface F: $z = z(x)$. To each point A of M we assign the point of the hyperplane $z = 0$ with coordinates $p_1 = \partial z/\partial x_1, \ldots, p_n = \partial z/\partial x_n$. This mapping is said to be *normal.* Let $\overline{M}$ denote the projection of the set M onto the hyperplane $z = 0$, and let $\widetilde{M}$ denote the normal image of M. Then the areas of the sets $\widetilde{M}$ and $\overline{M}$ are equal. Indeed, by Eq. (1),

$$\int_{\widetilde{M}} dp_1\, dp_2 \ldots dp_n = \int_{\overline{M}} \det(z_{ij})\, dx_1 \ldots dx_n = \int_{\overline{M}} dx_1 \ldots dx_n.$$

The concept of a normal mapping can be defined for a general convex hypersurface (nonregular). Then the quantities p_k are the slopes of the support hyperplanes (see subsection 1, Sec. 5).

To avoid introducing new notation, we shall understand by the hypersurface F that part of it is situated under the hyperplane σ: $z = h$. Construct the cone V with vertex O projecting the domain T_σ (Fig. 4). The normal image of the hypersurface F covers the

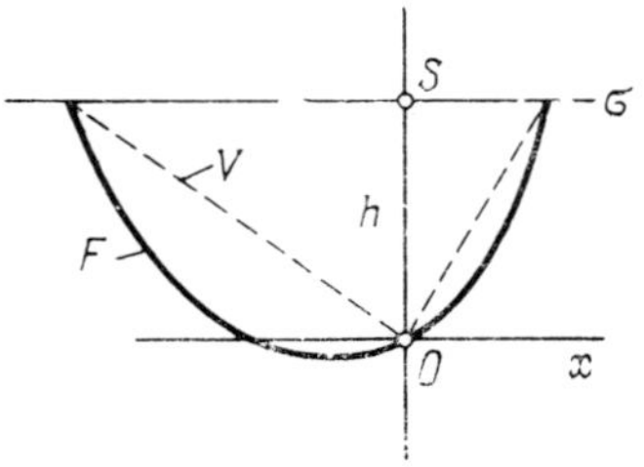

FIG. 4

normal image of the cone V because there exists a hyperplane tangent to F. The normal image of the cone V covers the normal image of the cone with vertex O which projects the hypersphere of smallest radius with center S that contains the body T_σ. Hence, the area of the normal image of F is no less than $(h/r)^n \varkappa_n$, where $\varkappa_n$ is the volume of the n-dimensional unit sphere. The area of the projection of F onto the hyperplane $z = 0$ is equal to the area of T_σ and therefore is no greater than $r^n \varkappa_n$. Since the area of the normal image of F is equal to the area of the projection of F onto the hyperplane $z = 0$, we obtain the inequality

$$\frac{(h/r)^n \varkappa_n}{r^n \varkappa_n} = (h/r^2)^n \leqslant 1,$$

i.e., $h/r^2 \leqslant 1$.

Let us now estimate the ratio h/r^2 from below. The hypersurface lying below the hyperplane σ is a bowl. Let us estimate its depth H (Fig. 5). The point A is outside the body T_σ and hence at a distance no less than $r/n^{3/2}$ from the point S (subsection 1). The point B is inside the body T_σ and hence at a distance no greater than r from the point S. From this it follows that

$$H = h \frac{AB}{AS} = h \left(1 + \frac{BS}{AS}\right) \leqslant h\,(1 + n^{3/2}).$$

On the negative semi-axis z, we fix a point Ω at a distance $2H$ from the point S (Fig. 6). Let us construct the cone V with vertex Ω projecting the hypersphere of radius $r/n^{3/2}$ with center at S.

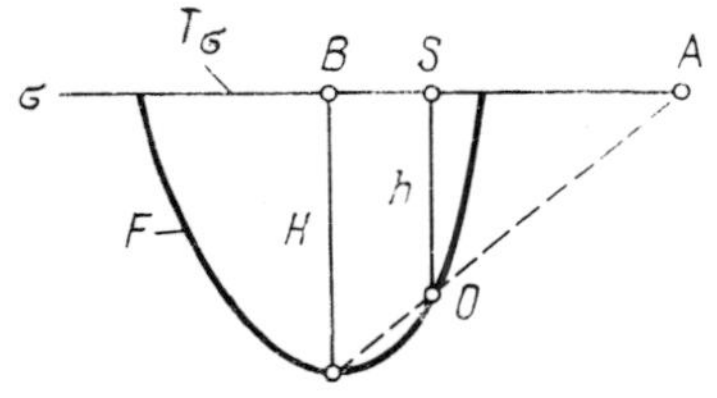

FIG. 5

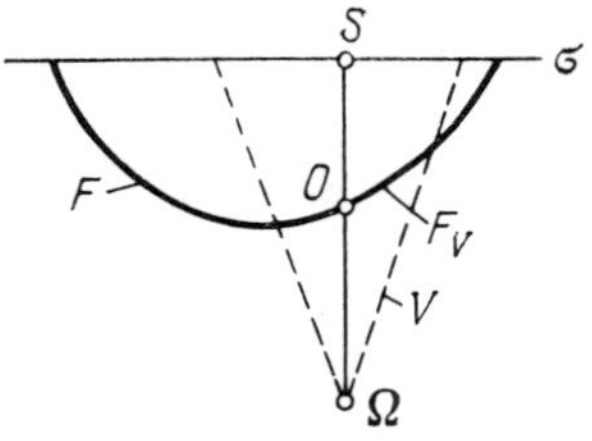

FIG. 6

Denote by F_V the part of F inside the cone V. The area of the normal image of the hypersurface F_V is no greater than the area of the normal image of the cone V, i.e., no greater than $\varkappa_n\ (2Hn^{3/2}/r)^n$. The projection of F_V onto the hyperplane $z = 0$ covers the hypersphere of radius $r/2n^{3/2}$ with center at O. Therefore, the area of the projection is no less than $\varkappa_n(r/2n^{3/2})^n$. From this we obtain the inequality

$$\frac{\varkappa_n(2Hn^{3/2}/r)^n}{\varkappa\ (r/2n^{3/2})} = \left(\frac{4Hn^3}{r^2}\right)^n \geqslant 1,$$

or, since $H \leqslant h(1 + n^{3/2})$,

$$\frac{h}{r^2} \geqslant \frac{1}{4n^3\,(1 + n^{3/2})}.$$

This completes the proof of the assertion.

4. *A priori* estimates for affine hyperspheres. Let $z(x)$ denote a convex solution of the equation

$$\det\,(z_{ij}) = 1 \tag{1}$$

in a convex domain G of diameter $D \leqslant 2$. Suppose that, on the boundary of G, the solution satisfies the boundary condition $z = 1$. Then for the first derivatives of z, where $z \leqslant 1/2$, there exists everywhere in G an estimate

$$|\,z_\alpha\,| \leqslant c_1,$$

where c_1 is a constant depending only on n.

Let $\bar{A}'$ denote an arbitrary point of the domain G and let A and A' denote the points on the hypersurface F: $z = z(x)$ and the hyperplane σ: $z = 1$, respectively, such that A projects into the point $\bar{A}'$ (Fig. 7). Denote by h' the length of the segment AA'.

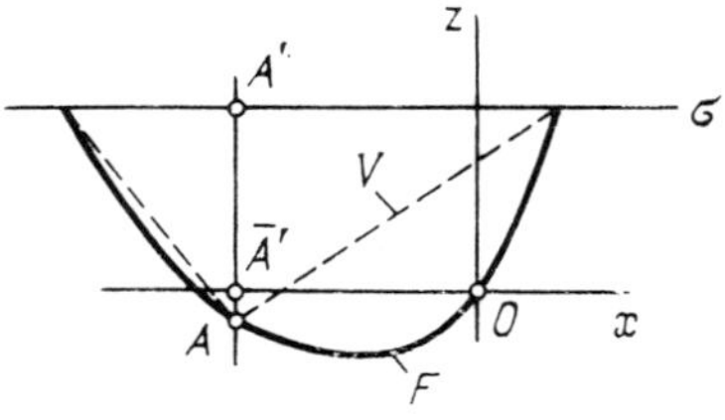

FIG. 7

Construct the cone V with vertex A projecting the edge of the hypersurface F. The area of the normal image of the cone V is no greater than the area of the normal image of the hypersurface F and hence is no greater than the area of the domain G. The area of G is no greater than $\varkappa_n D^n$.

The normal image of the cone V is a convex domain. Its boundary is defined by the equation

$$x = \frac{\xi h'}{H(\xi)},$$

where $H(\xi)$ is the distance of the support hyperplane of the domain G with exterior normal ξ from the point $\overline{A}'$. If the distance of the point $\overline{A}'$ from the boundary of the domain G is equal to ρ', then a point at a distance h'/ρ' from the coordinate origin O belongs to the normal image of the cone V. On the other hand, the normal image of the cone V covers the hypersphere of radius h'/D. Since the normal image of the cone is a convex domain, it contains a circular cone with height h'/ρ' and base of radius h'/D. It follows that the area (volume) of the normal image of the cone V is no less than

$$\frac{1}{n}\frac{h'}{\rho'}\left(\frac{h'}{D}\right)^{n-1}\varkappa_{n-1},$$

where $\varkappa_{n-1}$ is the volume of the $(n-1)$-dimensional unit sphere. Since the area of the domain G is no greater than $\varkappa_n D^n$, we obtain the inequality

$$\frac{1}{n}\frac{h'}{\rho'}\left(\frac{h'}{D}\right)^{n-1}\varkappa_{n-1} \leqslant \varkappa_n D^n.$$

Noting that now $|z_\alpha|_A \leqslant h'/\rho'$, we have for $z = 1/2$

$$|z_\alpha| \leqslant \frac{h'}{\rho'} \leqslant \frac{n\varkappa_n}{\varkappa_{n-1}} D^{2n-1}2^{n-1} \leqslant \frac{n\varkappa_n}{\varkappa_{n-1}} 2^{3n-2}.$$

This completes the proof of the assertion.

We shall show now that, for $z \leqslant 1/4$, the solution $z(x)$ of Eq. (1) admits an absolute estimate

$$| z_{\alpha\beta} | \leqslant c_2,$$

where c_2 is a constant depending only on n.

Let G_1 denote the subdomain of the domain G in which $z \leqslant 1/2$ and let G_2 denote the subdomain of the domain G_1 in which $z \leqslant 1/4$. It has been proved that in the domain G_1 we have $|z_\alpha| \leqslant c_1$, where c_1 is a constant depending only on n. Let x denote an arbitrary point of the domain G_2. We estimate the distance $\rho\ (x)$ from this point to the boundary of the domain G_1. The first derivatives of $z(x)$ are no greater than c_1 in the domain G_1; $z(x) \leqslant 1/4$ at the point x, and $z = 1/2$ on the boundary of the domain G_1. It follows that $\rho\ (x) \geqslant 1/4c_1$. Thus, we have a positive lower estimate, depending only on n, for the distances from points of G_2 to the boundary of G_1.

By Theorem 2 (Sec. 5), the second derivatives of the solution of Eq. (1) at an interior point of the domain G_1 admit an estimate depending only on the maximum of the absolute values of the first derivatives and the distance from the boundary of the domain G_1. Since in the domain G_2 the distances of the points from the boundary of G_1 are uniformly bounded from below by the positive number $(1/4c_1)$, the estimate for the second derivatives, for $z \leqslant 1/4$, depends only on c_1 and hence only on n. The assertion is proved.

5. Proof of the main theorem on improper convex affine hyperspheres. We shall prove that an improper convex affine hypersphere is an elliptic paraboloid. Let F denote an improper convex affine hypersphere defined by a convex function $z(x)$ such that

$$\det\ (z_{ij}) = 1. \tag{1}$$

Without loss of generality, the coordinate origin O is assumed to belong to F. On the semi-axis $z > 0$, consider a point S at a distance h from the origin O and the central section σ (see subsection 2) passing through the point S. We denote by T_σ the intersection of the body T (bounded by the hypersphere F) and the hyperplane σ. Let $\overline{T}_\sigma$ denote the projection of T_σ onto the hyperplane $z = 0$. On the semi-axis $z > 0$, let S' denote a point at a distance $h/4$ from the origin O and let σ' denote a hyperplane passing through S' and parallel to σ. Let $T_{\sigma'}$ denote the intersection of σ' and T and let $\overline{T}_{\sigma'}$ denote the projection of $T_{\sigma'}$ onto the hyperplane $z = 0$. The body $\overline{T}_{\sigma'}$ contains the body $\overline{T}_{\sigma/4}$ obtained from $\overline{T}_{\sigma'}$ by the homothetic transformation with respect to the center O with coefficient of homothety $1/4$. Suppose that

$$z = p_1x_1 + p_2x_2 + \dots + p_nx_n + h$$

is the equation of the hyperplane σ.

Perform the affine transformation

$$\begin{aligned} x_1' &= x_1,\ x_2' = x_2,\ \dots,\ x_n' = x_n, \\ z' &= z - p_1x_1 - \dots - p_nx_n \end{aligned}$$

of the hypersphere F. Under this transformation, the hyperplane σ is mapped into the hyperplane $z = h$.

We construct the ellipsoid of miminum volume with center at O that contains the body $\overline{T}_\sigma$. Subject the hyperplane $z = 0$ to an equiaffine transformation that maps the ellipsoid E into a hypersphere. Let r be the radius of this hypersphere. Setting $z' = z$, we extend the equiaffine transformation to the entire space. In a suitably chosen coordinate system, this equiaffine transformation is given by the formulas

$$z' = z,\ x_1' = \lambda_1x_1,\ x_2' = \lambda_2x_2,\ \dots,\ x_n' = \lambda_nx_n.$$

Let us now perform the affine transformation

$$z' = \frac{z}{h},\quad x_1' = \frac{x_1}{r},$$

$$x_2' = \frac{x_2}{r},\dots,\ x_n' = \frac{x_n}{r}.$$

Let $\bar{F}$ denote the affine hypersphere obtained from F as a result of these three affine transformations. Let $\bar{z}(x)$ denote the function which defines $\bar{F}$. The function $\bar{z}(x)$ satisfies the equation

$$\det(\bar{z}_{ij}) = \text{const} > 0.$$

Since $\lambda_1 \lambda_2 \ldots \lambda_n = 1$ and the ratio h/r^2 is bounded by absolute constants (see subsection 2), the constant in the equation for $\bar{z}$ is also bounded by an absolute constant (depending only on n). Without loss of generality, we may assume this constant to be equal to 1.

The second derivatives for the hypersurfaces F and $\bar{F}$ at corresponding points are connected by

$$\bar{z}_{\alpha\alpha} = \left(\frac{1}{\lambda_\alpha}\right)^2 \frac{r^2}{h} z_{\alpha\alpha}. \tag{*}$$

The affine transformation of F into $\bar{F}$ maps the point O into itself. Therefore,

$$\bar{z}_{\alpha\alpha}\big|_O = \left(\frac{1}{\lambda_\alpha}\right)^2 \frac{r^2}{h} z_{\alpha\alpha}\big|_O.$$

Taking into consideration the estimates for $\bar{z}_{\alpha\alpha}$ (see subsection 4), we conclude that the coefficients of the equiaffine transformation admit an absolute lower estimate. Since $\lambda_1 \lambda_2 \ldots \lambda_n = 1$, we get an absolute upper estimate for them.

The affine transformation of F into $\bar{F}$ maps the domain $\bar{T}_{\sigma'}$ into a domain G'. In this domain, $z \leqslant 1/4$. Therefore, there exists an absolute estimate for the derivatives $\bar{z}_{\alpha\alpha}$ (see subsection 4). Now, taking into account the absolute estimates for λ_α and r^2/h, we conclude from (*) that there exist absolute estimates for $z_{\alpha\alpha}$ in the domain $\bar{T}_{\sigma'}$.

As $h \to \infty$, the radius $r \to \infty$. Since there exist absolute estimates for λ_α independent of h, the domain $\bar{T}_\sigma$ extends over the whole hyperplane $z = 0$ as $h \to \infty$. We now draw an important conclusion: the hypersurface F is projected onto the whole hyperplane $z = 0$. The relation between $\overline{T}_\sigma$ and $\overline{T}_{\sigma'}$ ($\overline{T}_{\sigma/4} \subset \overline{T}_{\sigma'}$) enables us to conclude that, as $h \to \infty$, the domain $\overline{T}_{\sigma'}$ extends over the whole hyperplane $z = 0$. Since in the domain $\overline{T}_{\sigma'}$ there exists an absolute estimate for $z_{\alpha\alpha}$ which is independent of h, this estimate is valid over the whole hyperplane $z = 0$:

$$| z_{\alpha\alpha} | \leqslant c.$$

Because of Eq. (1), we obtain an estimate from below for $z(x)$;

$$z_{\alpha\alpha} \geqslant \frac{1}{c^{n-1}}.$$

Let us consider now the Riemannian metric defined by the metric tensor

$$g_{\alpha\beta} = z_{\alpha\beta}.$$

Since

$$\sum_{\alpha,\beta} g_{\alpha\beta}\, dx_\alpha\, dx_\beta \geqslant \frac{1}{c^{n-1}} \sum_\alpha dx_\alpha^2.$$

this metric is a complete one. Hence, as was shown by E. Calabi [10], the solution $z(x)$ of Eq. (1) is a second-degree polynomial, i.e., the hypersphere F is an elliptic paraboloid. This completes the proof of the theorem.

In conclusion, we should mention the following: The hypotheses of the theorem put no regularity conditions on the function $z(x)$ defining the hypersphere except the natural requirement of twice-differentiability. But the proof of the theorem assumes that $z(x)$ is of class C^4 (subsection 4), and Calabi's theorem, which completes the proof, requires existence of fifth derivatives of $z(x)$. However, the regularity requirements were not included in the statement of the theorem because the twice-differentiability of the solution of Eq. (1) implies strict convexity of the affine hypersphere and by Theorem 5, Sec. 5, such a solution is analytic.

BIBLIOGRAPHY

1. H. Minkowski, Volumen und Oberfläche, *Math. Ann.*, **57** (1903).
2. T. Bonnesen und W. Fenchel, *Theorie der konvexen Körper*, Ergebnisse der Math., 1934.
3. A. D. Aleksandrov, On the theory of mixed volumes of convex bodies, *Mat. Sb.*, **2**, 5, 6 (1937); **3**, 1, 2 (1938) (Russian and German summaries).
4. H. Lewy, On differential geometry in the large I, *Trans. Amer. Mat. Soc.*, **43**, 2 (1938).
5. C. Miranda, Su un problema di Minkowski, *Rend. Sem. Mat. Roma*, **3** (1939).
6. L. Nirenberg, The Weyl and Minkowski problems in differential geometry in the large, *Comm. Pure Appl. Math.*, **6** (1953).
7. D. Hilbert, *Grundzüge einer allgemeinen Theorie der linearen Integralgleichungen*, Leipzig–Berlin, 1915.
8. A. V. Pogorelov, *Monge-Ampère Equations of Elliptic Type*, Izdat. Khar'kov. Gos. Univ., Khar'kov, 1960; English translation, Noordhoff, Groningen, 1964.
9. K. Jörgens, Uber die Lösungen der Differentialgleichung $rt - s^2 = 1$, *Math. Ann.*, **127** (1954).
10. E. Calabi, Improper affine hyperspheres of convex type and a

generalization of a theorem by K. Jörgens, *Michigan Math. J.*, **5**, 2 (1958).
11. C. Miranda, *Partial Differential Equations of Elliptic Type*, Springer–Verlag, Berlin, Heidelberg, New York, 1970.
12. H. O. Cordes, Uber die erste Randwertaufgabe bei quasilinear Differentialgleichungen zweier Ordnung in mehr als zwei Variablen, *Math. Ann.*, **131** (1956).
13. A. V. Pogorelov, An *a priori* estimate for the principal radii of curvature of a closed convex hypersurface in terms of its Gaussian curvature, *Dokl. Akad. Nauk SSSR*, **181**, 4 (1968); English translation, *Soviet Math. Dokl.*, **9**, 4 (1968).
14. A. V. Pogorelov, On a regular solution of the n-dimensional Minkowski Problem, *Dokl. Akad. Nauk SSSR*, **199**, 4 (1971); English translation, *Soviet Math. Dokl.*, **12**, 4 (1971).
15. A. V. Pogorelov, An *a priori* estimate for the principal radii of curvature of a closed convex hypersurface in terms of their mean values, *Dokl. Akad. Nauk SSSR*, **186**, 3 (1969), English translation, *Soviet Math. Dokl.*, **10**, 3 (1969).
16. A. V. Pogorelov, Existence of a closed convex hypersurface with a prescribed function of the principal radii of curvature, *Dokl. Akad. Nauk SSSR*, **197**, 3 (1971); English translation, *Soviet Math. Dokl.*, **12**, 2 (1971).
17. A. V. Pogorelov, On regularity of generalized solutions of the equation $\det(\partial^2 u/\partial x_i \partial x_j) = \varphi(x_1, x_2, \ldots, x_n) > 0$, *Dokl. Akad. Nauk SSSR*, **200**, 3 (1971); English translation, *Soviet Math Dokl.*, **12**, 5 (1971).
18. A. V. Pogorelov, The Dirichlet problem for the n-dimensional analogue of the Monge-Ampère equation, *Dokl. Akad. Nauk SSSR*, **201**, 4 (1971); English translation; *Soviet Math. Dokl.*, **12**, 6 (1971).
19. A. V. Pogorelov, Improper convex affine hyperspheres, *Dokl. Akad. Nauk SSSR*, **202**, 5 (1972); English translation, *Soviet Math. Dokl.*, **13**, 1 (1972).